Nicolas Zehner
Envisaging Dataist Modernity

Science Studies

Nicolas Zehner (Dr.) is a postdoctoral researcher at the Collaborative Research Center 1265 "Re-Figuration of Spaces" at Technische Universität Berlin, as well as associated researcher at Weizenbaum Institute. He holds a PhD in Sociology from the University of Edinburgh. In the academic year 2021/2022, Zehner was a visiting research fellow at Harvard University's Program on Science, Technology and Society. His research investigates the co-constitutive relationship between scientific knowledge production and urban development with a particular focus on higher education institutions.

Nicolas Zehner

Envisaging Dataist Modernity

The Construction of Edinburgh's Innovation Apparatus

[transcript]

This book is a revised version of the author's dissertation in Sociology at the University of Edinburgh.

The publication of this book was generously funded by the CRC 1265 "Re-Figuration of Spaces" at Technical University Berlin.

Bibliographic information published by the Deutsche Nationalbibliothek
The Deutsche Nationalbibliothek lists this publication in the Deutsche Nationalbibliografie; detailed bibliographic data are available in the Internet at https://dnb.dnb.de

Hermannstraße 26 | D-33602 Bielefeld | live@transcript-verlag.de

Cover design: Maria Arndt
Cover illustration: Anthony Ashley on Pixabay
Printing: Majuskel Medienproduktion GmbH, Wetzlar
https://doi.org/10.14361/9783839475294
Print-ISBN: 978-3-8376-7529-0 | PDF-ISBN: 978-3-8394-7529-4
ISSN of series: 2703-1543 | eISSN of series: 2703-1551

Printed on permanent acid-free text paper.

Contents

Acknowledgments

This book would not have been possible without my two supervisors, Nick and Nathan. Meeting for the first time in September 2018, we had only a very limited idea of what this project could become. My interest in cities and imagined futures was matched by their respective expertise in urban and economic sociology. What followed was an incredibly stimulating collaboration between two scholars who – until that point – had not co-supervised together and one Berliner who set out to discover Edinburgh. Nick, I am grateful for your intellectual free spiritedness, your patience and constant encouragement. Nathan, your critical and provocative thinking has been instrumental in exploring my academic identity and finding my place in this (sometimes) intimidating world. The two of you have shaped my intellectual and personal development tremendously for which I am deeply thankful.

If people on the ground had not granted me access to their world, this work would not have come to fruition. This includes interview partners and informants, who shared their knowledge, concerns and hopes about the city region and its futures. This also includes scholars who took a leap of faith and invited me to contribute. Participating in Liz McFall's and AWED's "Closes and Opens: A History of Edinburgh's Futures" gave me the opportunity to better understand how Edinburgh's pasts, presents and futures are interconnected and constantly overlapping. Contributing a chapter to Morgan Currie's, Jeremy Knox's and Callum McGregor's "Data Justice and the Right to the City" allowed me to put my thoughts down on paper very early on. Working with Suzanne Ewing provided me insight into the intricacies of scientific knowledge organisation. Lastly, I want to thank the Scottish Graduate School of Social Science for awarding me a ESRC Studentship. The financial and professional support has been invaluable and I am forever grateful.

This research project was profoundly shaped by my time at Harvard's "Program on Science, Technology and Society". I am deeply grateful to Sheila for inviting me to spend a year in Cambridge and her persistent encouragement to ask the big questions. My work benefitted enormously from countless discussions with other STS fellows, attending classes at the Kennedy School and the Graduate School of Design and presenting my research to various audiences. Thinking deeply and boldly about how science and society co-produce each other has enriched my thinking profoundly. I will be eternally grateful for this opportunity.

Finally, I want to express my deep gratitude to my friends and family. Without their constant encouragement and support I would not be the person I am today. Lots of love to Konsti, Felix, Tim, Korbi, Jack, Rix, Julius, Bruno, Karl, Maxi, Zac and Allie, Gabriel, José Tomás and Fran, Lillan, Olivier, Pauli, Dr Peters, Yasmine, Pablo, Nils and my sister Alisa. Most importantly, I am blessed with an amazing life, which I owe to my parents. Special thanks to Tomi who is one of the most inspiring and supporting persons I have ever met.

Lay summary

Drawing on the sociology of expectations and urban studies, this book explores who gets to imagine what kinds of urban-regional futures. In doing so, it investigates the emergence and early implementation of a large, multi-stakeholder urban-regional development project, the Edinburgh and South-East Scotland City Region Deal (CRD). The CRD constitutes a £1.3bn infrastructure investment by the Scottish and UK governments and local partners over fifteen years and is designed to transform the city region into the "Data Capital of Europe" (DCoE). In investigating empirically this project by drawing on documentary research, 50 elite interviews and participant observations, the study asks questions such as: Who is involved in envisioning desirable city-regional futures? Why is it that data-driven innovation and technology are identified as future engines of societal prosperity? What role do universities play in transforming urban-regional environments? The book's main finding is that the agency of actors to creatively change existing ideas about the world according to their hopes for the future is unequally distributed. The thesis further shows that the future of the Edinburgh city region is being negotiated and enacted by technocratic elites – political leaders, tech entrepreneurs, higher education officials – at the intersection of science and politics. Specifically, it demonstrates the growing influence of universities in shaping urban-regional development. Putting forward this argument, the concept of 'dataist modernity' is developed to characterise the contemporary relationship between science, technology and urban planning.

Abstract

This book investigates the process by which urban planning agents imagine futures and the sociotechnical conditions which help establish the credibility of their expectations. Aiming to shed light on the practice of city-making, it focuses on how urban planning agents co-produce specific urban-regional sociotechnical imaginaries. Based on the assumption that city-regional innovation processes are inherently future-oriented and imply a profound reordering of society, the book pays close attention to who gets to construct and diffuse imaginaries of desirable futures. Drawing on documentary research, 50 interviews and ethnographic observations, this study empirically engages with the implementation of a contemporary urban-regional redevelopment project, the Edinburgh and South-East Scotland City Region Deal (CRD). A £1.3bn infrastructure investment by the Scottish and UK governments and local partners over fifteen years, it is designed to accelerate data-driven innovation and inclusive growth. Officially signed in August 2018, the CRD represents a unique case study for at least two reasons: first, it constitutes the first higher education-driven city region deal in the UK. Second, it aligns a cluster of considerably powerful actors – political leaders, higher education officials, tech entrepreneurs – around one specific sociotechnical imaginary: establishing the city region as the "Data Capital of Europe". Conceptually rooted at the intersection of Science and Technology Studies (STS) and Urban Studies, this book argues that future-making in the Edinburgh city region is characterised by an unequal distribution of projective agency and driven by an entrepreneurial class at the nexus of science and politics. Shedding light on the complex interplay of expectations and expertise in urban-regional development, it demonstrates the growing influence of global higher education institutions such as the University of Edinburgh in shaping desirable futures. More specifically, the concept of "dataist modernity" is developed to characterise a form of modernisation that is rooted in an ontology of knowability and overestimates the power of data in solving complex social problems. By crafting this line of argument, this study extends co-productionist STS on 'smart urbanism' to take fuller account of the intricate relationship between technoscience and urban-regional planning.

List of Figures

List of Abbreviations

ACF	Advanced Computing Facility
AI	Artificial Intelligence
ANT	Actor-Network-Theory
ARIA	Advanced Research and Invention Agency
ATMs	Automated Teller Machine
BEIS	Department for Business, Energy and Industrial Strategy
BIS	Department for Business, Innovation and Skills
BLE	Bluetooth Low Energy
CGI	Canadian IT and business consulting firm
CIAM	Congrès International d'Architecture Moderne
COO	Chief Operating Officer
CORFO	Corporación de Fomento de la Producción
COSLA	Convention of Scottish Local Authorities
CPD	Continuing Professional Development
CRD	City Region Deal
DCoE	Data Capital of Europe
DDI	Data-Driven Innovation
DECC	Department of Energy and Climate Change
EFI	Edinburgh Futures Institute
EIDF	Edinburgh International Data Facility
EMEA	Europe, the Middle East and Africa
ENTEL	National Telecommunications Enterprise
EPCC	Edinburgh Parallel Computing Centre
FCDC	Future Cities Demonstrator Competition

FPTP	First-Past-The-Post
GDP	Gross Domestic Product
GIA	Gross Internal Area
GoF	Gang of Four
GVA	Gross Value Added
HEIs	Higher Education Institutions
HESA	Higher Education Statistics Agency
IoT	Internet of Things
IPCC	Intergovernmental Panel on Climate Change
KBUD	Knowledge-based Urban Development
LoRaWAN	Long Range Wide Area Network
MBS	Mohammed bin Salman
MIPIM	Real Estate Exhibition
MIT	Massachusetts Institute of Technology
MOOC	Massive Open Online Course
NGO	Non-Governmental Organisation
NHS	National Health Service
NYU	New York University
QMU	Queen Margaret University
R&D	Research and Development
RDAs	Regional Development Agencies
REC	Regional Enterprise Council
SIA	Science and Innovation Audit
SIGMA	Science in General Managment
SSK	Sociology of Scientific Knowledge
STIs	Sociotechnical imaginaries
STS	Science and Technology Studies
STV	Single-Transferable-Vote
TRADE	Talent, Research, Adoption, Data, Entrepreneurship
TSB	Technology Strategy Board
UCL	University College London
UK	United Kingdom
UKRI	United Kingdom Research and Innovation

UN	United Nations
UNCOVER	Usher Network for Covid-19 Evidence Reviews
UNEP	United Nations Environment Programme
VSM	Viable System Model
WEF	World Economic Forum

Chapter 1: From Cannes to Edinburgh – Entering Dataist Modernity

1. Imagining urban-regional futures

> "Sometimes you play a single word in your mind for a while and you wake up at three o'clock in the morning. Thank you. Here we are. You see a solution to a problem. You see a solution to the way ahead." (Local council leader, 2020)

> "We have the city we deserve. We have the city everybody takes part in and arrives at a conclusion on." (Former council leader, 2020)

Urban-regional utopias are more alive than ever. The two images below show perhaps the most extraordinary – and absurd – example of how science and technology are mobilised to imagine urban-regional futures. Part of NEOM[1], a $500bn (!) urban-regional development project by the Saudi Arabian government, "The Line" is destined to spearhead the nation's transition from oil-dependent into technology and innovation-driven economic futures. At 200 metres wide, 170 kilometres long and 500 metres above sea level (roughly the height of the Empire State Building), "The Line" is intended to be a hyper-connected, AI-powered, clean-energy city that will be home to nine million people. In January 2021, Saudi Arabia's crown prince, Mohammed bin Salman (MBS),

1 More information can be found here: https://www.neom.com/en-us

announced the launch of "The Line" claiming it represented "a civilisational revolution that puts humans first, based on a radical change in urban planning".[2]

Figure 1: "The Line", left: Drone footage from construction site[3]; right: Prototype | Sources: Ot Sky and NEOM

At the time of writing, bulldozers and excavators have begun carving out the contours of Saudi Arabia's utopian dreams. Interestingly, the idea of a linear city is far from new. Instead, Nereim (2022) points out that "The Line" "carries echoes of an idea originated in the 1960s by Superstudio, an Italian architectural collective, for a structure so enormous it would wrap around the entire planet" (p. 5). Ironically, this structure – Superstudio's "Continuous Monument" (see Figures 2 and 3) – "was never a real building proposal [...] [but] intended as a critique of excessive urbanization and of the modernist megaprojects then in vogue" (Nereim, 2022, p. 6). Confronted with MBS's vision of the future in a New York Times interview, Gian Piero Frassinelli – one of the remaining Superstudio members – comments: "Seeing the dystopias of your own

2 See the full press release here: https://www.youtube.com/watch?v=xLN2Vu1iels

3 More footage here: https://www.youtube.com/watch?v=HfiWYKTgRxI

imagination being created is not the best thing you could wish for" (Imam, 2021).

Figure 2: Drawing from the "Continuous Monument" | Source: Archivio Superstudio

"The Line" constitutes an evocative case for several reasons. First, it shows how close utopia and dystopia can be. Superstudio's dystopian view and critique of total urbanisation was taken up 50 years later by MBS and hailed as a model for desirable urban futures. Second, "The Line" offers an insight into the soul life of the most powerful groups in society. It demonstrates what is possible to imagine. It is capitalist imagination taken to its limit. It illuminates what happens when enormous economic and political capital – embodied in the Saudi Arabian government, prestigious architectural firms and consultancies – meet science fiction. Apparently fascinated by the aesthetic and culture of cyberpunk – a subgenre of sci-fi that typically depicts dark, tech-infused futures – MBS has tasked an army of designers, consultants, writers and technologists to transform fiction into reality (Nereim, 2022). To sum it up, "The Line" encapsulates three contemporary phenomena that increasingly shape economic action. First, the future functions as a mode of governance. That is, it becomes actionable via imaginaries and narratives as central decision-making mechanisms of the present. Second, cities and city regions represent crucial socio-economic spaces for realising desirable futures. Finally, science and technology constitute key instruments of the imagination, governing urban-regional futures.

Figure 3: Photomontage from the "Continuous Monument" series completed in 1970 | Source: Centre Pompidou, MNAM-CCI

This book attempts to grapple with the interplay of imagined futures, city regions and technoscience by exploring the ways in which urban planning agents imagine futures, and the sociotechnical conditions which help establish the credibility of their expectations. More specifically, it investigates the intricate relationship between technoscience and cities through the lens of urban-regional innovation. Doing so, it asks the following overarching research questions: How do technoscientific and sociopolitical orders co-produce the city region? Who is permitted to implement and hegemonise specific imaginaries of the data-fuelled city? Why is it that data-driven innovation has been identified as a key vehicle to imagine the *good*[4] society?

4 The idea of the "good" points to the normative and political nature of future imaginaries. Different actors have different understandings of the "good". Because they usually do not align, the conflict-laden idea of the "good society" – how society *ought* to develop – opens up space for sociological analysis.

2. Technoscience and urban-regional planning

Cities and city regions have become world-making (Amin and Thrift, 2016). They exist simultaneously in a variety of forms: as physical entities, as cultural imaginaries and as datafied extrapolations. Cities are profoundly temporal settings and actors themselves, aligning pasts, presents and futures. As Paul Virilio and Patrick Geddes aptly point out: the city is a "box full of speeds" (Virilio and Lotringer, 2007, p. 77) and "a drama in time" (Geddes, 1904, p. 107). The significance of the city – understood as both a "practico-material and architectural fact" and "a social reality" (Lefebvre, 1996, p. 103) – cannot only be described in philosophical terms, but is reflected in figures that depict the sociotechnical realities on the ground.

According to the United Nations (2018), an estimated 55 per cent of the world's population reside in urban areas. This number is projected to increase to 68 per cent by 2050. City regions drive economic growth. The World Bank (2022) reports that more than 80 per cent of global GDP is generated in cities. Furthermore, urban-regional areas play a key role in driving and fighting climate change. According to the UNEP (2022), cities are currently responsible for 75 per cent of global CO_2 emissions. It follows that they play a central role in mitigating the effects of the climate crisis. Cities are sites of multidimensional, overlapping inequalities, ranging from social and economic to spatial and digital inequality. The decline of manufacturing and subsequent rise of the "new economy" – centrally built around an information-based mode of production – has led to widening income gaps, growing inter-city disparities and increasing housing unaffordability (Nijman and Wei, 2020). Finally, the Covid-19 pandemic has reversed years of efforts in fighting urban poverty. The UN Habitat *World City Report* (2022) documents the emergence of "newly poor" people – "those who would have exited poverty in the absence of Covid-19 but are now projected to remain poor as well as those projected to fall into poverty" (p. 78) – as a result of the pandemic. The number of "newly poor" people was projected to rise to between 143 and 163 million people globally in 2021 (ibid.). In summary, these statistics underline the importance of paying close attention to who, how and what is (re)shaping, (re)forming and (re)configuring urban environments.

One response to these urban-regional challenges or "wicked problems" (Rittel and Webber, 1973) is to come up with a plan. In its most fundamental form, planning can be understood as "a way of conceptualizing space and time, and the possibilities that time offers space" (Abram and Weszkalnys, 2011, p. 3).

Planning implies the attempt to control futures. Planning represents efforts to deal with the "futurity paradox"; meaning, "the inevitable tension between the future-oriented nature of economic action and the 'uncalculability' of the uncertain future" (Braun, 2015, p. 368). Crucially, to plan means to govern. Planning is inherently political. Wildavsky (1973) aptly notes that "to plan is to make decisions that affect others. Planners are presidents, ministers, bureaucrats, party leaders, scientists, entrepreneurs – anybody whose acts have large future consequences" (p. 133). This book traces the emergence and early implementation of a plan to transform the Edinburgh city region. During a business conference in Cannes in 2016, a senior official from the University of Edinburgh and a senior executive from the City of Edinburgh Council began to envision a city region driven by data and innovation. Engaging with key planning agents from the fields of government, education and business, this study asks: Who is planning urban-regional futures? What are some of the ontological and epistemological assumptions undergirding the planning process? What happens if plans fail?

The governance of city-regional futures is increasingly performed with the help of technoscience (Thrift, 2014; van Dijck, 2014; Farías and Blok, 2017; Zukin, 2020). One imaginary that is particularly powerful when it comes to reshaping city regions pertains to the notion of the "smart city" (Greenfield, 2013; Kitchin, 2014b; Coletta *et al.*, 2018; Lindner and Meissner, 2019; Karvonen, 2020). Similar to other catch-all terms, the concept of the "smart city" is marked by its nebulous nature and is not reducible to a single meaning. Recognising the interpretive flexibility of the term, smart urbanism can be treated as a broad indicator for the increasing embedding of networked digital infrastructures into the urban fabric. At the heart of all efforts is data, the "material produced by abstracting the world into categories, measures and other representational forms [...] that constitute the building blocks from which information and knowledge are created" (Kitchin, 2014a, p. 2). One actor increasingly shaping smart city governance is universities and research institutions. According to UN Habitat's *Global Review of Smart City Governance Practices* (2022), "universities and research institutions are the second most active stakeholders after municipal governments [...] in both [the] planning and implementation phases" (p. 37) of smart city initiatives. This book sheds light on the University of Edinburgh – both as a site of knowledge production and as an urban planning agent – as a key player in the politics of city-regional future-making.

The irrepressible thirst for "smartness" spans from academia (Karvonen, 2020), to business (Sadowski and Bendor, 2018) and policy (Morozov and Bria, 2018). This research study critically investigates notions of the "smart city" in the context of thc Edinburgh city region. In doing so, it asks: What kinds of scientific knowledge are mobilised in the conceptualisation of the "smart"? How do grandiose and future-oriented imaginaries around data-driven innovation relate to existing ideas and imaginaries about the city region? What new concepts might we need to deal with an urban world that is increasingly embedded with machine learning and big data infrastructures?

3. Studying projective agency in the Edinburgh city region

The research problem of this project is concerned with projective agency in an urban planning setting. Projective agency denotes "the imaginative generation by actors of possible future trajectories of action, in which received structures of thought and action may be creatively reconfigured in relation to actors' hopes, fears, and desires for the future" (Emirbayer and Mische, 1998, p. 971). Projective agency constitutes the primary unit of analysis and is operationalised by use of Jasanoff's and Kim's (2015) notion of "sociotechnical imaginaries". Sociotechnical imaginaries are defined as "collectively held, institutionally stabilized, and publicly performed visions of desirable futures, animated by shared understandings of forms of social life and social order attainable through, and supportive of, advances in science and technology" (ibid., p. 4).

Examining projective agency in an urban planning setting, this study engages with the emergence and early implementation of a contemporary urban redevelopment project, the Edinburgh and South-East Scotland City Region Deal (henceforth CRD). The CRD represents a £1.3bn investment over fifteen years, which is designed to accelerate inclusive growth through the funding of infrastructure, skills and data-driven innovation. Taking the CRD as its primary empirical case study, this book investigates the ways in which urban planning agents imagine futures by examining the construction and diffusion of a specific urban-regional sociotechnical imaginary: establishing the Edinburgh city region as the "Data Capital of Europe" (DCoE).

Tackling this research problem, this study presents a sociological intervention into the sociotechnical process by which urban planning agents construct desirable futures. Doing so, it is theoretically and analytically rooted at the intersection of Science and Technology Studies (STS) (Beck, 1992; Gieryn, 1995; Scott, 1998; Jasanoff, 2004; Jasanoff and Kim, 2015; Pfotenhauer and Jasanoff, 2017b) and urban studies (Amin and Thrift, 2002; Brenner, 2004; Kitchin, 2016; Coletta *et al.*, 2018; Sadowski, 2020; Zukin, 2020). At this point, it makes sense to briefly examine how the highly interdisciplinary disciplines of STS and urban studies have interacted in the past.

Traditionally, the field of urban studies has neglected the co-constitutive role of science and technology and the city (Dierig, Lachmund and Mendelsohn, 2003; Farías and Blok, 2017). Farías (2009) points out that urban studies has conceptualised cities as 1) definite spatial environments and ecological niches, 2) as capitalist politico-economic formations, and 3) as production sites of cultural practices. In other words, the city was approached as a spatial formation within which human communities settle down as result of ecological processes, such as competition, succession and invasion (Burgess, 1925; Park, 1936); as a politico-economic actor shaped by and shaping capitalist economies (Weber, 1958; Lefebvre, 1991) and as "culture", involving urban practices such as contestation, resistance and protection from the stimuli of the metropolis (Simmel, 1903; Jacobs, 1961).

What is missing from these accounts is an appreciation of and engagement with the mutually shaping roles of science and technology in cities and city regions. Traditional accounts in urban thinking have treated "the city" as a bounded unit and stable object, thereby 1) neither problematising science and technology as constitutive of what it is that defines the urban, 2) nor paying attention to how scientific knowledge-producing practices and urban politics co-produce each other (Farías and Blok, 2017). It is against this backdrop that scholars in STS, geography and urban planning started to make interventions in urban studies. STS accounts of the urban range from studying the city as a technological artifact (e.g., Aibar and Bijker, 1997), to examining the urban built environment as technological objects (e.g., Gieryn, 2002), to recasting urban ontology altogether (Farias and Bender, 2010; McFarlane, 2011). In addition to that, there is a rich body of literature on networked urbanism, which explores the co-constitutive role of technology and the city by studying cities as sociotechnical assemblages and investigating the complex relations between infrastructure networks and urban life (Mitchell, 1995; Graham and Marvin, 2001; Kitchin and Dodge, 2011; Townsend, 2013). Finally, there is a significant body of work that acknowledges the world-shaping role of knowledge and knowledge infrastructures in driving urban-regional development (Etzkowitz and Leydesdorff, 1995; Castells, 2010; Carrillo *et al.*, 2014).

This book builds on these existing accounts and subscribes to the "co-productionist" strand of STS (Jasanoff, 2004). The idiom of co-production suggests "that the ways in which we know and represent the world (both nature and society) are inseparable from the ways in which we choose to live in it" (Jasanoff, 2004, p. 2). Thus, the analytical framework of co-production stresses that technoscientific and sociopolitical orders are mutually constitutive and dependent. This particular strand of STS can be understood as emerging from a specific intellectual trajectory. Building on the Edinburgh School's "Sociology of Scientific Knowledge (SSK)" (Bloor, 1991) and the Paris School's "Actor-Network Theory (ANT)" (Callon, 1984; Latour, 2005), co-production seeks to avoid both social and technoscientific determinism.

In other words, it does not treat science and technology as wholly independent variables but rather as forces that influence, and are influenced by, politics. It acknowledges epistemic hybridity and multiple forms of agency but notes that SSK ignores materiality and ANT drifts into depoliticising power by downplaying structure and history. Jasanoff (2015) points out that materialist accounts such as ANT and assemblage urbanism are "too distributive, too promiscuous in attributing cause and agency" [thereby] "risking a form of moral nihilism, making all actions and agents seem equally empowered, or disempowered, and therefore equally responsible or irresponsible for the networks within which they function" (pp. 16–17).

In order to further delineate "co-production" from ANT-inspired approaches of the urban such as "assemblage urbanism" (Farias and Bender, 2010; McFarlane, 2011; Blok and Farias, 2016), it makes sense to zoom in on the notion of "symmetry". Scholars such as Ignacio Farías and Colin McFarlane advance a symmetrical understanding of human and non-human actors. They put forward a notion of materiality that is *active*. Agency is equally distributed between actors and actants. Co-production, on the other hand, adopts a *passive* understanding of materiality, restricting agency to individual and social actors. Materials are only bearers of normative visions. Thus, co-production "is symmetrical in that it calls attention to the social dimensions of cognitive commitments and understandings, while at the same time underscoring the epistemic and material correlates of social formations" (Jasanoff, 2004, p. 3).

Simply put, the idiom of co-production grapples with the observation that *we make the worlds we wish to live in*. This book studies the co-production of technoscientific and socio-political orders by tracing the *emergence* of Edinburgh's innovation apparatus. In doing so, it grasps urban-regional innovation through the concept of "sociotechnical imaginaries" (STIs), which can be understood as "mechanisms of co-production". That is, STIs are co-productionist in at least two salient ways. They appeal to "an integrated examination of societies across scales and domains of representation and social practice" and are "a symmetrical inquiry that asks how knowledge and materiality are co-produced with social values and normativity" (Jasanoff and Simmet, 2021, p. 2). To recap, this book treats the city and city region as a highly intriguing site for studying the production of desirable futures with science and technology. It extends co-productionist STS on "smart urbanism" by interrogating how science and technology become enmeshed in producing collective visions of the "good society".

3.1. Research questions

Addressing the research problem at hand, the book is built around the following three sets of research questions and sub-questions:

1. What role does scientific expertise play in shaping contemporary imaginaries of the datafied city (region)?
    - Why are particular kinds of urban-regional knowledge given importance while other strands of knowledge are side-lined?
2. How are data-driven fictional futures practiced, individually and collectively?

- Which competing imaginaries of the modern university are at play when reconceptualising the Edinburgh city region as the "Data Capital of Europe"?

3. What implications can be drawn from the study of projective agency in the Edinburgh city region for the relationship between technoscience and urban-regional planning in 21st century societies?
    - What role does technoscience play in re-centring the social, i.e., aligning social orders around new or different understandings of urban-regional sociality and the social good?

Exploring these research questions, this project pursues three aims. The first aim is to make a substantive contribution to existing research at the intersection of co-productionist STS and urban studies by illuminating cities, more generally, and higher education institutions (HEIs), more specifically, as central sites for the evolution and expansion of STIs. The second aim of this book is to contribute towards a better understanding of the notion of smart urbanism by shedding light on a city region which, so far, has received only limited academic attention (Karvonen, Cugurullo and Caprotti, 2019; Currie, Knox and McGregor, 2022). In *Who runs Edinburgh?*, David McCrone (2022) provides a detailed analysis of Edinburgh's political economy. Examining transformations in post-war politics, the city's cultural power and recent planning controversies, he presents a fascinating history of Edinburgh's pasts and presents. Crucially, he does not address Edinburgh's promised futures around data and technology, as inscribed in the emergence of the CRD. This book picks up where he left off and illuminates *who shapes Edinburgh's futures*. Finally, this study aims to provide a conceptual framework that historicises and contextualises previous attempts to manage urban-regional complexity with the help of technoscience against the background of Edinburgh's imagined futures.

3.2. Summary

This research project investigates the process by which urban planning agents imagine futures and the sociotechnical conditions which help establish the credibility of their expectations. Aiming to shed light on the practice of city-making, it focuses on how urban planning agents co-produce specific urban-regional STIs. Based on the assumption that city-regional innovation processes are inherently future-oriented and imply a profound reordering of society, this work pays close attention to who gets to construct and diffuse imaginaries of desirable futures.

This study draws on documentary research, 50 interviews and ethnographic observations to empirically engage with the implementation of a contemporary urban-regional redevelopment project, the Edinburgh and South-East Scotland CRD. As a £1.3bn infrastructure investment by the Scottish and UK governments and local partners over fifteen years, the CRD is designed to accelerate data-driven innovation and inclusive growth. Officially signed in August 2018, the CRD represents a unique case study for at least two reasons. First, it constitutes the first higher education-driven CRD in the UK. Second, it aligns a cluster of considerably powerful actors – political leaders, higher education officials, tech entrepreneurs – around one specific sociotechnical imaginary: establishing the city region as the DCoE.

Conceptually rooted at the intersection of STS and urban studies, this research argues that future-making in the Edinburgh city region is characterised by an unequal distribution of projective agency, and driven by an entrepreneurial class[5] at the nexus of science and politics. Shedding light on the complex interplay of expectations and scientific expertise in urban-regional development, this study demonstrates the growing influence of global HEIs such as the University of Edinburgh in shaping desirable futures. More specifically, the concept of "dataist modernity" is developed to characterise a form of modernisation that is rooted in an ontology of knowability which overestimates the power of data in solving complex social problems. Rooted in the ideology of "dataism" (van Dijck, 2014; Harari, 2017; Fourcade and Gordon, 2020; Davies, 2022) – which can be roughly defined as the deep-seated belief in the objectivity of quantification and its application in the most complex domains of social life –, the concept of "dataist modernity" attempts to interrogate the role of technoscience in contemporary urban planning.

3.3. Outline of book

This work traces the emergence and early implementation of Edinburgh's innovation apparatus. To this end, it is centrally structured around Chapters 4–7, which highlight different units of analysis: the apparatus (Chapter 4), the actors (Chapter 5), the institution (Chapter 6) and the bigger picture (Chapter 7). This main part of the book is complemented by a chapter outlining the theoretical framework (Chapter 2), a chapter presenting the methodological approach (Chapter 3), and a concluding chapter (Chapter 8) summarising the main findings and offering some meta-level reflections. Figure 4 gives an overview of the structure of the project.

5 In what follows, I adopt a loose definition of entrepreneurial class. Closely related to what Schwartz (2014) calls "professional-administrative class" and Kitchin *et al.* (2017) describe as "smart city epistemic community", the notion of "entrepreneurial class" refers to powerful individuals with similar sociocultural backgrounds who share the belief in a technology-driven and risk-taking approach to urban-regional development.

Figure 4: Book structure | Source: Own illustration

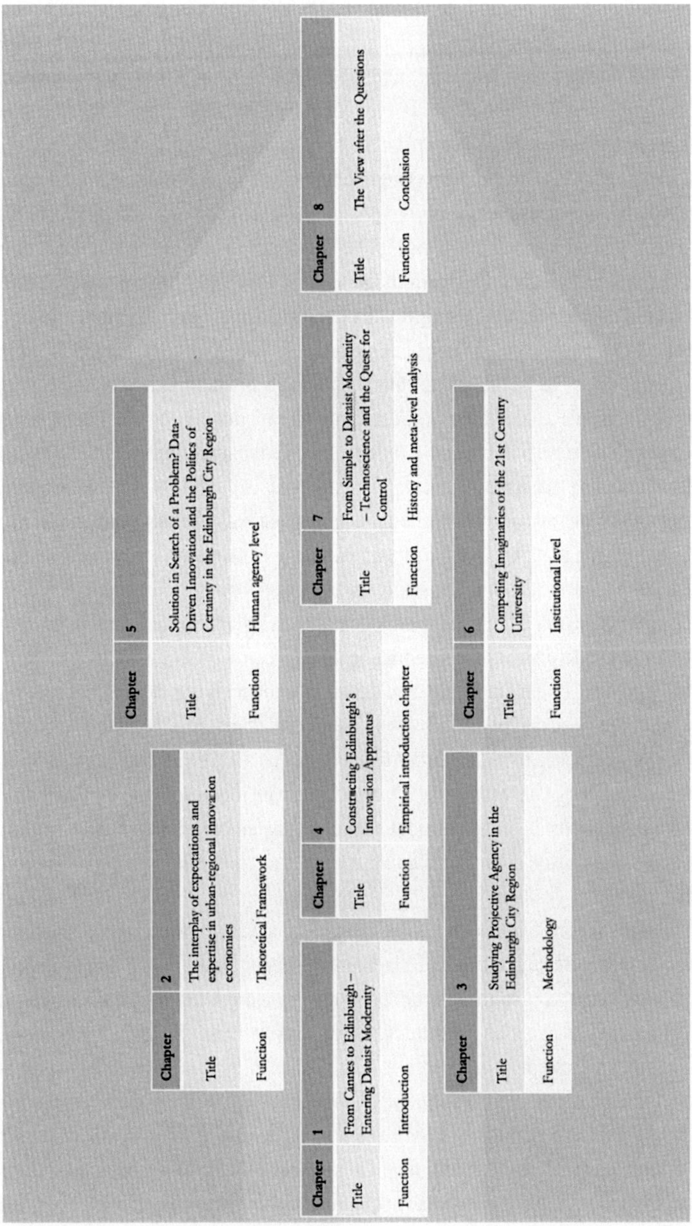

Chapter 2 introduces the theoretical framework undergirding the book. Academic literature at the intersection of the sociology of imagination and science (Beck, 1992; Gieryn, 1995; Jasanoff, 2004; Jasanoff and Kim, 2015; Beckert, 2016; Pfotenhauer and Jasanoff, 2017b) provides a useful lens to study data-driven urban-regionalism. Advancing this line of argument, the chapter reviews relevant theoretical debates and empirical discussions in sociology, political economy, STS and urban geography along three broad thematic strands: 1) capitalism's temporal order and decision-making, 2) data-driven innovation and the city, and 3) the relationship between scientific expertise and democracy. Three key insights form the basis of the forthcoming analysis. First, expectations, narratives and imaginaries provide highly promising units of analysis for a better understanding of the dynamics of capitalist societies. Second, empirical studies in STS highlight that the politics and science of the future are closely intertwined, and also demonstrate the importance of investigating the intricate relationship between scientific expertise and democratic legitimacy. Finally, scholarly contributions at the intersection of STS and urban studies conceptualise cities as sociotechnical assemblages, thereby emphasising the distribution of agency among human and non-human actors as well as depicting the urban as being in a constant state of becoming.

Chapter 3 paints the picture of my research journey. More informal and reflexive in style, it provides a chronological account of the research conducted for this project. It begins by introducing the underlying empirical case study and elaborates on my initial interest in the relationship between expectation formation and urban planning. Afterwards, it discusses the methodological approach guiding the research. This includes an outline of the methods used and an explanation of the rationale and challenges inherent of data generation and analysis. Finally, ethical considerations and risks that shaped the research are discussed.

Chapter 4 constitutes the first empirical chapter and lays the analytical foundation for the remainder of the book. Tracing how the CRD emerged from a conversation in a pub in Southern France to a significant £1.3bn investment aiming to transform the Edinburgh city region into the DCoE, this chapter demonstrates how imagination is turned into social practice. Combining Jasanoff's and Kim's (2015) concept of STIs with the analytical vocabularies of urban assemblage thinking, it argues that Edinburgh's innovation apparatus can be understood as a "promissory assemblage" (Färber, 2019), which aligns actors, organisations, discourses and geographical spaces around the promise of eternal economic growth. Focussing on three analytical interlinkages –

scales, identities and agencies – Edinburgh's drive for innovation is described by introducing key agencies and drawing out empirical tensions that form the basis for subsequent chapters.

Building on insights from Chapter 4 – particularly on three overlapping policy changes which provided space to reimagine the relationship between science, cities, politics and innovation – Chapter 5 explores the role of human agency in shaping Edinburgh's innovation apparatus. More specifically, it engages with the role of the so-called "Gang of Four" (GoF) – a group of four powerful university officials – in structuring power relationships and mediating the epistemology-politics intersection in the context of data-driven urban-regionalism. Seeking to understand how it is possible to frame complex social issues such as slow growth and inequality as problems of innovation, this chapter argues that reconceptualising the Edinburgh city region as a relevant space of innovation and positioning the University of Edinburgh as its key projective agent required a process of apoliticisation. Defining the latter as the *orchestration of temporal ruptures*, it demonstrates how the GoF mobilised language, location and data science in order to construct "an imaginary of a fixed and scripted, indeed, inevitable, future" (Schiølin, 2020, p. 545).

Chapter 6 ascends the analytical and organisational ladder and explores the role of powerful HEIs such as Edinburgh University in 21st century urban-regional development. To this end, it asks: How does the university, as an institution but also as a site of knowledge production, affect necessary transitions of the 21st century? What are the conditions that allow HEIs to move from humble servants of the public good to powerful future-makers? Addressing these research questions, the chapter argues that the University of Edinburgh's shift from traditional self-imaginations such as the "ivory tower" to more globally, networked imaginaries, embodied in the notion of the "entrepreneurial institution". Thus implying a reconfiguration of technoscientific and sociopolitical orders. Importantly, this transition does not unfold in an isomorphic fashion but is shaped by the institution's attempt to transform into an innovation agent while living up to its civic responsibilities. This mode of "institutional eclecticism", however, calls into question existing lines of accountability and necessitates a critical rethinking of the role of the modern university in urban-regional development.

Chapter 7 provides a meta-level reflection on the relationship between technoscience and planning by mobilising two historical examples: Le Corbusier's high modernism from the 1920s and Salvador Allende's cybernetic socialism from the 1970s. Building on these two historical reference points, the chapter develops the notion of the "neo-scientisation of urban-regional planning" against the background of what Beck, Giddens and Lash (1994) call "reflexive modernisation". In doing so, it contextualises the emergence of Edinburgh's innovation apparatus with earlier attempts to steer society and situates this research's empirical findings within a contemporary moment referred to as "dataist modernity". The latter is defined as a form of modernisation which is rooted in an ontology of knowability and which overestimates the power of data in solving complex social problems. The following research questions guide the chapter: What role does technoscience play in re-centring the social? That is, aligning social orders around new or different understandings of sociality and the social good. How have "scientific tools" of ordering society changed over time? How does the emergence of big data and artificial intelligence – key ingredients for imagining smart futures in the Edinburgh city region – relate to Corbusier's scientific management and Beer's management cybernetics? Tackling these questions, Chapter 7 argues that HEIs such as the University of Edinburgh become key agents in managing contemporary urban-regional complexity. What used to be the master planner in the early 20th century and the planning state in 1960s and 1970s, are now "sub-political" organisations such as universities that assume – knowingly or unknowingly – responsibility in shaping smart urban-regional futures.

Finally, Chapter 8 – The View after the Questions – summarises the main argument and provides a discussion of the book's contributions and limitations. In addition, the three units of analysis from Chapters 5–7 – the actors, the institution and the bigger picture – are briefly revisited. In the process, some meta-level conclusions are drawn about the relationship between science and the city and future research avenues are identified.

Chapter 2: The Interplay of Expectations and Expertise in Urban-Regional Innovation Economies

1. Introduction

This book investigates the interplay of expectations and expertise in urban-regional innovation projects. Specifically, it studies the process by which urban planning agents imagine futures and the sociotechnical conditions that help establish the credibility of their expectations. Aiming to shed light on the practice of city-making, this study examines how urban planning agents co-produce and diffuse precise urban-regional sociotechnical imaginaries. To that end, it engages with the following four main research questions:

1) How are urban planning agents developing expectations about the future?
2) What is the form and content of data-driven urban futures?
3) How are fictional urban futures practiced, individually and collectively, by whom and when?
4) What role does scientific expertise play in shaping contemporary imaginaries of the datafied city (region)?

This study is concerned with projective agency in an urban planning setting. Projective agency denotes "the imaginative generation by actors of possible future trajectories of action, in which received structures of thought and action may be creatively reconfigured in relation to actors' hopes, fears, and desires for the future" (Emirbayer and Mische, 1998, p. 971). The latter constitutes the primary unit of analysis and is operationalised by use of Jasanoff's and Kim's (2009, 2015) notion of "sociotechnical imaginaries" (STIs). STIs are defined as "collectively held, institutionally stabilized, and publicly performed visions of desirable futures, animated by shared understandings of forms of social life and social order attainable through, and supportive of, advances in science and technology" (ibid., p. 4). Put differently, this book investigates the ways in which urban planning agents imagine futures by examining the construction and diffusion of a specific urban-regional STI – establishing the Edinburgh city region as the "Data Capital of Europe" (DCoE).

The purpose of this chapter is to set the scene. In this chapter, I suggest that academic literature at the intersection of the sociology of imagination and science (Beck, 1992; Jasanoff, 2004; Beckert, 2013a; Jasanoff and Kim, 2015; Beckert, 2016; Eyal, 2019) is applicable to a new kind of data-driven urban-regionalism. I begin by reviewing relevant theoretical debates and empirical discussions in sociology, political economy, science and technology studies (STS) and urban geography that grapple with three thematic strands: 1) capitalism's temporal order and decision-making, 2) data-driven innovation and the city, and 3) the relationship between scientific expertise and democracy. At the end of each section, I will reflect on existing shortcomings and provide conceptual starting points for the underlying study. Finally, concluding remarks follow.

2. Theoretical framework

2.1. Capitalism's temporal order and decision-making

2.1.1 Economic decision-making and temporality

This research project focuses on the temporal dynamics of economic action. In seeking to gain a better understanding of the inner workings of capitalist economies' mode of reproduction, it is essential to grasp on what basis actors make decisions in economic contexts. Accordingly, one central research question prevails: How do actors manage and navigate the tension between capitalist economies' inherent future orientation and the incalculability of an uncertain future? In this section, I will critically review existing sociological contributions, which deal with the time dimension in economic decision-making and the role of uncertainty.

The temporalities of capitalism

The evolution of capitalism as the dominant social formation of economic action was accompanied by changes in the temporal orientations of actors, and an altered understanding of the future as a temporal category and conceptual space (Rammstedt, 1975; Adam, 2010; Beckert, 2016). This change in temporal orientation is rooted in the evolutionary development of social structures – materialised through events such as the Reformation in the 16th century and the Enlightenment movement of the 17th century – from simple, undifferentiated societies to functionally differentiated societies of high modernity (Rammstedt, 1975).

While former social structures are characterised by an "occasional" time consciousness, that is, experiences of time which only differentiate between a "now" and a "not now", the later are marked by a linear time consciousness with an open and uncertain future. The experience of time becomes one of continuous movement and acceleration (Rosa, 2013). It follows that the genesis of the capitalist order is shaped by changes in socio-temporal structures and the subsequent enlargement of time horizons into an unknown economic future. Hence, economic actors constantly strive for a "different and 'better' future that holds the promise of increased consumption, higher profits and new lifestyles" (Beckert, 2016, p. 26).

As a result, scholars describe capitalism's inherent logic of escalation as "dynamic disequilibrium" (Beckert, 2009), "restlessness" (Sewell, 2008) or "dynamic stabilisation" (Rosa, 2015). Hartmut Rosa points out that modern capitalist society can only function according to a mode of dynamic stabilisation; that is, "it systematically requires growth, innovation and acceleration for its structural reproduction and to maintain its socio-economic and institutional status quo" (Rosa, 2015, p. 283). It follows that the capitalist system is marked by a "growth totalitarianism" (ibid., p. 76); that is, "regardless of the actual volume of national output, it can never reach a conclusion" (ibid.). To sum it up, capitalist dynamics are characterised by a logic of escalation and an open and uncertain future.

The role of uncertainty

Acknowledging that the future in modern capitalist societies is recognised as open, the question of how economic actors deal with the uncertainty of the future becomes highly relevant. Braun (2015) calls this the "futurity paradox", "the inevitable tension between the future-oriented nature of economic action and the 'uncalculability' of the uncertain future" (p. 368). How do economic actors make decisions under conditions of uncertainty? Tension between the unknowability of the uncertain future and social actors' constant attempts to know the future, portrays one central research puzzle of this research project.

Key to an investigation into the relationship between economic action and uncertainty is a clear conceptualisation of uncertainty. Frank Knight's distinction between risk and uncertainty provides a helpful starting point. Knight (1921) points out that "if we are to understand the workings of the economic system we must examine the meaning and significance of uncertainty" (p. 199). Initially distinguishing between three types of probability situations – "a priori probability", "statistical probability" and "estimates" (pp. 224–225) – Knight later condenses these three types into two categories, which he calls "risk" and "uncertainty". Risk denotes those situations in which "the distribution of the outcome in a group of instances is known (either through calculation a priori or from statistics of past experience)" (Knight, 1921, p. 233). In situations marked by uncertainty, on the other hand, "it is impossible to form a group of instances, because the situation dealt with is in a high degree unique" (ibid.).

In summary, risk is measurable and uncertainty is not. Assuming that uncertainty depicts a prevailing condition in economic decision-making situations, it is essential to critically review how existing sociological accounts explain the role of the future and uncertainty in economic action.

2.1.2 Expectations and imagined futures

If economic action in modern capitalist societies is shaped by changes in cognitive orientations through new temporal dispositions, then present action cannot be explained through the past alone but must also be influenced by perceptions of the future. These future perceptions are called expectations. The discipline that has most rigorously examined expectations in the social sciences is economics. Here, expectations are conceptualised through rational expectations theory and, its corollary, the efficient market hypothesis (Muth, 1961; Fama, 1965; Lucas, 1972). Expectations are seen to be determined through rational calculation. In other words, based on the statistical distribution of past events, economic actors can make correct predictions of the future. This approach has been heavily criticised by scholars within and beyond economics (Keynes, 1964; Dequech, 2006; Hong and Sullivan, 2013). Most relevant for the context of this study, is the fact that economics disregards the fundamental uncertainty of economic action as a result of the openness of the future, and therefore does not account for the contingent nature of expectations (Beckert, 2016).

Modern sociological contributions, examining the role of the future in economic action emphasise the importance of norms, institutions, structures of social networks or imagination (Parsons, 1951; DiMaggio and Powell, 1983; Granovetter, 1985; Beckert, 2013, 2016). Examining the impact of uncertainty on expectations, existing accounts differ concerning the degree of openness and creative potential they attach to the role of expectations in propelling capitalist dynamics. Parsons (1951), for instance, emphasises expectations as a stabilising mechanism, as a means to resolve the problem of "double contingency". Scholars such as Granovetter and Swedberg (2011), on the other hand, highlight social structures. Challenging the artificial separation between economics and sociology, they point to the social constitution and structuring of markets ("embeddedness") as an explanatory factor for the outcome of economic decisions. Others emphasise that isomorphism and path dependency cause economic outcomes (DiMaggio and Powell, 1983; Mahoney, 2000). Contrary to accounts which explain economic action primarily by reference to the past, Jens Beckert (2013, 2016) advocates a more open assessment of the impact of uncertainty on expectations. Adopting a pragmatist standpoint (Dewey, 1922), Beckert recognises the contingency of expectations and the creativity entailed in the imagining of future situations.

Finally, there are a plethora of scholarly approaches within STS that study the future. They all point to the future as constitutive of the present, they emphasise the performative force of future-oriented statements, representations and practices, and they reveal a general concern about how to steer and shape technoscience in positive directions (Konrad *et al.*, 2017). Two concepts are particularly relevant for this research project: expectations and imaginaries. Building on a variety of case studies, the sociology of expectations emphasises the collective and performative character of expectations in technological innovations (Lente, 1993; Brown and Michael, 2003; Brown, Rip and Van Lente, 2003; Borup *et al.*, 2006; Pollock and Williams, 2018). Borup *et al.* (2006) define technological expectations as "real-time representations of future technological situations and capabilities" (p. 286).

STS researchers also recognise images and imaginaries as highly significant in all instantiations of technoscience (Latour, 1990; Fujimura, 2003; Jasanoff and Kim, 2009; Ezrahi, 2012; Jasanoff, 2015a; Karvonen, 2020). One concept that has gained particular prominence in recent years is Jasanoff's and Kim's (2009, 2015) notion of STIs: "collectively held, institutionally stabilized, and publicly performed visions of desirable futures, animated by shared understandings of forms of social life and social order attainable through, and supportive of, advances in science and technology" (Jasanoff and Kim, 2015, p. 4). In what follows, I will elaborate in more detail on expectations and imaginaries as useful units of analysis to study projective agency in an urban planning context.

Sociology of expectations

The thriving field of the sociology of expectations is strongly shaped by the economic sociology of imagination (Bronk, 2009; Beckert, 2013b, 2016; Beckert and Bronk, 2018), and a further subfield of STS which focuses on the role of expectations in science and technology (Lente, 1993; Brown and Michael, 2003; Brown, Rip and Van Lente, 2003; Borup *et al.*, 2006; Pollock and Williams, 2018).

The role of uncertainty in economic action takes centre stage in this line of thought. Uncertainty is understood as "the character of situations in which agents cannot anticipate the outcome of a decision and cannot assign probabilities to the outcome" (Beckert, 1996, p. 804). In other words, since dynamic capitalist economies are characterised by relentless innovation and novelty, they exhibit an indeterminacy which cannot be reduced to measurable risk (Knight, 1921). Challenging the notion of rational expectations, Beckert (2016) concludes that "actors making decisions on investments, innovations, credits, or consumption develop imaginaries on how the future will look and how the decisions they make will influence outcomes" (p. vii). Put differently, scholars affiliated with this research agenda investigate how imaginaries and narratives interact with calculative devices to structure expectations and beliefs in the economy (Beckert and Bronk, 2018).

This line of argument adopts a relational, pragmatist perspective on agency. According to Emirbayer and Mische (1998), human agency can be defined as "the temporally constructed engagement by actors of different structural environments—the temporal-relational contexts of action—which, through the interplay of habit, imagination, and judgment, both reproduces and transforms those structures in interactive response to the problems posed by changing historical situations" (p. 970). Particularly relevant for the purposes of this study is the authors' emphasis on the projective element of agency. Projectivity "encompasses the imaginative generation by actors of possible future trajectories of action, in which received structures of thought and action may be creatively reconfigured in relation to actors' hopes, fears, and desires for the future" (ibid., p. 971). Noting that sociological theory and research have mostly[1] neglected the analysis of the future, Mische (2009, 2014) urges scholars to revive the notion of projectivity as a promising tool for social analysis. She points out that attention should refocus "on the open, indeterminate, 'polythetic' perception of the field from the point of view of the

1 The subfield of "utopian studies", most prominently represented by Ruth Levitas (2013), denotes a notable exception.

actor surveying the future in terms of multiple possibilities, as opposed to the 'monothetic' view of the actor (or observer) who interprets the decision after it has already been taken" (Mische, 2009, p. 696).

In line with scholars such as Mische (2009) who call for researching futures in action, Jens Beckert (2013) developed the notion of "fictional expectations" – the inhabitation in the mind of an imaginary future state of the world – which assertively foregrounds the role of imagination as a tool for coping with radical uncertainty in economic decision-making. Challenging rational expectations theory and existing sociological approaches, which emphasise the role of social macrostructures for shaping actors' expectations, Beckert opens up a new way of understanding the micro foundations of the dynamics of the economy by advancing a "non-teleological theory of action that brings the creativity of actors and the contingency of the future into the foreground" (Beckert, 2013b, p. 235). In doing so, he advances the imagined-futures approach to economic sociology, which suggests that the investigation of expectations about the future contributes towards a better understanding of the dynamic nature of capitalist societies.

The role of imaginaries of the future can be studied in various spheres of the economy. Examples include investment, consumption or innovation. Particularly relevant for this piece of research are sociological analyses that engage with the formation and role of expectations in innovation. According to Beckert and Ergen (2021), research into imagined futures in innovation studies can be divided into: 1) accounts that emphasise the history of STIs, 2) studies that explore the micropolitics of expectations in technological development, and 3) empirical research which focuses on how the future is opened up and uncertainty (deliberately) created.

Imaginaries of science, technology and society

Imaginations of the future are central to organisation and economic activity. They are important for understanding projects that rely on a high degree of projective agency (Emirbayer and Mische, 1998). Imaginaries of the future justify new investments in science and technology, and mobilise and legitimise individual and organisational action. They are produced through a variety of social practices, and are embedded in cultural and institutional conditions. During the last few decades, STS researchers have increasingly focused on the concept of imaginaries as a promising lens to gain a better understanding of the interplay between science, technology and society (McNeil *et al.*, 2017).

According to McNeil *et al.* (2017), STS work with imaginaries can be divided into three broad clusters: 1) cultures, communities and practices, 2) nations, institutions, and policies, and 3) bodies, subjects and differences. The strand of STS research most relevant for this PhD project deals with imaginaries in the context of nations, institutions and policies. Here, Jasanoff's and Kim's (2009) comparative study of US and South Korean orientations towards civil nuclear power technology illustrates the most influential contribution. Stressing the cultural and institutional embeddedness of divergent future expectations, the authors introduce the concept of STIs (Jasanoff and Kim, 2009, p. 120) as defined in the introduction.

Building on late-twentieth century sociopolitical theorists such as Benedict Anderson (1983), Charles Taylor (2004) and Cornelius Castoriadis (1998), Jasanoff and Kim (2015) seek to pull together collective imaginations such as Anderson's (1983) "imagined communities" with STS's focus on hybrid networks and assemblages. Crucially, STIs are "not limited to nation-states [...] but can be articulated and propagated by other organized groups, such as corporations, social movements, and professional societies" (Jasanoff, 2015b, p. 4). Examples of applying this concept include Elta Smith's (2015) work on corporate imaginaries of biotechnology, Stephens et al.'s (2013) study on institutional imaginaries of publics in stem cell banking, or Sadowski's and Bendor's (2018) contribution on corporate narratives in the drive for smart urbanism.

2.1.3 Shortcomings and conceptual starting points

Current sociological and STS accounts, examining the role of future orientation for social action, provide excellent contributions to a better understanding of the dynamics of capitalist innovation economies. However, there are a range of shortcomings which this research project seeks to address. First, the sociology of expectations is not yet capable of satisfactorily answering theoretically how – that is, through which channels, techniques and social networks – expectations emerge and why they become credible or not. Whilst acknowledging the inherently contingent character of expectations and future visions, more research needs to be conducted with regard to the different dimensions of fictional expectations. This includes exploring

> the source of expectations (i.e. individual actor, professional actor, corporate actor), their intended audience (the self vs. others), and the tools with which different expectations compete for the attention of the audience

(political power, prestige, financial resources, rhetorical skills, etc. (Guseva and Mooney, 2018, p. 339).

This study contributes theoretically by focusing on two crucial elements of projective agency in organisations: first, the form and content of urban futures, and, second, the process of imagination itself. A second shortcoming concerns the adoption of uncertainty as a conceptual starting point for a sociological analysis of innovation economies. How uncertain is the future really? Put differently, does uncertainty – as an analytical category – merely serve as a justification for delineating the research agenda of the sociology of expectations from economics, or does it constitute a prevailing condition in economic decision situations? It appears that economic actors already move and act in prestructured environments. In other words, the problem of uncertainty seems to exist only in very specific, "Schumpeterian" economic contexts. Such as during processes of innovation or situations associated with change or crisis. Beckert (2016) argues that it is exactly those situations (of novelty, change and crisis) which are most closely related to the dynamics of capitalism.

What about banal, mundane and everyday situations? Do they characterise the dynamics of capitalism to a lesser degree? This book, by studying projective agency in an urban planning context, contributes towards a better understanding of the role of uncertainty in economic decision-making. Rather than discussing the role of expectations in the "politics of *un*certainty", it seems appropriate to explore the "politics of certainty" by examining how grandiose claims such as transforming the Edinburgh city region into the DCoE relate to the mundane realities of regional economic development. This tension will be ad dressed in Chapter 5.

Third, the relationship between uncertainty and novelty must be examined in more detail. Beckert (2016), following Shackle (1979), argues that the contingency of expectations offers the possibility of creative change in the economy, through choices based on imaginaries of future states of the world. Expectations are shaped by the ability of humans "to see and do things in a novel way" (Dequech, 1999, p. 422). How far do actors simply comply with existing norms and socio-economic structures? Under what circumstances are economic actors able to deviate from existing norms and structures and create something "new"? This tension constitutes one of the key issues this book investigates. Is it enough to claim that the indeterminacy of the future naturally leads to the creative force of economic action?

Applying this line of thought to the empirical case of Edinburgh, one logical question is: How does "epistemic culture" (Knorr-Cetina, 1999) shape urban planning agents' ability to imagine novel future states? This book seeks to shed light on the cultural and institutional framing that allows certain urban visions to emerge and others not to. After all, one would assume that changing institutions leads to changing expectations. This study puts particular emphasis on examining the social foundations of fictional expectations in urban governance – as exemplified by opportunity structures, cultural frames, institutions, networks, cognitive devices, mass media and past experiences.

Finally, STS work on imaginaries and recent contributions on STIs provide an extremely valuable analytical lens for the examination of urban-regional innovation projects. While Jasanoff and Kim (2015) have extended and developed the concept of STIs – particularly regarding scope – cities and city regions as well as higher education remain understudied as empirical objects of analysis. Sadowski and Bendor (2018) point out that "the majority of research on STIs has focused on the state as the central site for the evolution and expansion of STIs, specifically via national policies, regulations, and institutions" (p. 4). This book sheds light on higher education as a central site for the evolution and expansion of urban-regional STIs. Studying the Edinburgh CRD – a higher education-driven and urban-regional economic development project – I examine how higher education officials, together with city governments, leaders and planners, shape and extend the resulting DCoE imaginary.

To summarise this section, several conceptual starting points emerge. First, it becomes clear that temporality serves as a promising analytical lens for the study of the social cause and effect of economic phenomena. Temporal structures and horizons represent one, if not the, systematic link between actor and system perspectives (Rosa, 2013). Examining the temporal order of capitalism generates intriguing research paths with regard to the everyday reproduction of contemporary Western societies. Second, futures – understood as temporal categories and conceptual spaces – have been neglected as objects of study within the social sciences. John Urry (2016) points out that "future visions have powerful consequences and social science needs to be central in disentangling, debating and delivering those futures" (p. 7). It follows that social science in general, and sociology and STS in particular, must reclaim the terrain of future studies and assume responsibility when it comes to articulating *social* futures.

In this book, I study projective agency by conceptualising the DCoE vision as an urban-regional STI (Jasanoff and Kim, 2015, p. 4). Adopting the concept of STIs as the key underlying epistemological prism necessarily requires justification. There is no doubt that the notion of STIs shows strong similarities with existing, classic theorisations of social imaginaries in the social sciences. Examples include Benedict Anderson's (1983) famous notion of "imagined political communities" as a way of understanding the nature of nationalism, Arjun Appadurai's (1990) use of imaginaries in the context of globalisation or Charles Taylor's (2003) concept of "social imaginaries", which investigates how the durable social structures of Western modernity are sustained through collective practices, stories and ideas. Crucially, what is missing from these classic accounts is an explicit acknowledgement and detailed investigation of Western modernity's two most salient forces: science and technology. Recent contributions in the sociology of expectations such as Beckert's (2013b, 2016) work on fictional expectations or accounts on the performative character of expectations in technological innovations (Lente, 1993; Borup *et al.*, 2006) inspire social scientists to observe acts of future-making in capitalist economies. But in order to do so it is vital to also attend to the politics and democratic features that shape innovation economies.

The notion of STIs allows us to do both: it is at once descriptive of attainable futures and prescriptive of the kinds of futures that ought to be attained (Jasanoff, Kim and Sperling, 2006). In order to illustrate the usefulness of the concept, it makes sense to briefly state what STIs are *not*. Unlike policy agendas (Kingdon, 2014), STIs are less explicit and goal-driven. While sharing the properties of being collective and systemic with the notion of discourse (Hajer, 1995), STIs put stronger emphasis on the orchestration of materiality in the development of technoscience. Finally, unlike political ideology, STIs are moveable and open to reaching alternative futures. To sum up, the concept of STIs constitutes a helpful epistemological lens to develop an understanding of the co-constitutive role of technoscience and the city. While it certainly has its own shortcomings – for example, its tendency to focus on the meta- and policy-level at the expense of a detailed examination of the inner workings of material infrastructures – the notion of STIs provides an invaluable analytical perspective for the purposes of the forthcoming analysis.

2.2 Data-driven innovation and the city

"But a city is more than a place in space, it is a drama in time" (Geddes, 1904, p. 107)

Cities are now the defining organisational units of our time (Amin and Thrift, 2016). Cities not only provide citizens and organisations with space and capacity to organise, as meta-organisations and configurations, they can also be considered drivers of organisational action in their own right. Cities are organisational actors and organising contexts for grand challenges and diverse communities of people and organisations. Late capitalist economies are increasingly shaped by the idea of data-driven, information-led "smart" cities. Digital technologies are now central to how powerful urban stakeholders – including governmental, higher education and entrepreneurial bodies – imagine and bring urban futures into being (Karvonen, Cugurullo and Caprotti, 2019).

The progression from analogue to digital operations (digitisation) has led to a second kind of digital transformation: datafication (Mayer-Schönberger and Cukier, 2013). Datafication comprises "the masses of digital traces left by people and technologies in online spaces and the proliferation of advanced tools for the integration, analysis, and visualisation of data patterns for purposes of decision making and commercialisation" (Flyverbom, Deibert and Matten, 2019, p. 6). In this section, I will critically engage with literature at the intersection of urban and technological development, as exemplified through the smart city agenda. Key questions include: How does datafication affect the strategic capacity of urban governance arrangements? What role do technological imaginations play in times where big societal utopias are absent? Most importantly, what characteristics of the city influence how specific imaginaries of "smart" are enacted? Tackling these questions, I will first review the literature dealing with the relationship between capitalist dynamics and urban life, and then engage with work that focuses on the intersection of technological and urban development, most clearly exemplified by the smart city agenda.

2.2.1 Capitalism, modernity and the city

The growth of the modern city is deeply intertwined with the emergence of industrial capitalism during the 18th and 19th century. Industrialisation not only revolutionised the way people live, as reflected in the process of urbanisation, but also changed the way people think about society (Stevenson, 2003). Sociology, as an academic discipline, emerged in tandem with urbanisation and industrialisation and took cities – the places where the consequences and contradictions of societal change were most evident and most profoundly experienced – as their natural objects of analysis. Particularly relevant for this study are accounts that approach modernism through a temporal lens. Keeping in mind the overall research interest – projective agency in the urban development – I will briefly review academic perspectives that problematise the relationship between acceleration and the city.

Central to early sociological thought, examining the impact of industrialisation on society, was the rural-urban dichotomy – the implicit and explicit comparison between the city and the countryside. Most prominently, Ferdinand Tönnies (2012) introduced the concepts *Gemeinschaft* and *Gesellschaft*. The former signifying intimate, more habitual and enduring social relationships in rural communities, and the latter describing contractual and secondary social relationships, as found in large industrial cities. Similar to Tönnies' notion of *Gesellschaft*, Berlin sociologist Georg Simmel (1903) characterises urban modernity as shaped by the increasing rapidity of things. Examining the impact of the "money economy" on the pace of life, Simmel (2004) notes that "the mere increase in the supply of money that one has in one's hand intensifies [...] the temptation to spend money, and in doing so promotes a greater turnover in commodities, an increase, acceleration and multiplication in economic conceptions" (p. 505). Describing the modern city as the embodiment of the fusion of money and speed, Simmel gives an idea of how and where capitalist imaginaries are formed (Simmel, 2004).

Importantly, he argues that the imperatives of the market create modern urban dwellers who act instrumental, *blasé* and reserved (Simmel, 1903). Walter Benjamin (1999), another early theorist of the acceleration society, provides a more nuanced understanding of the impact of acceleration on urban life through his discussion of Baudelaire's figure of the *flâneur* (the casual observer of street-life). In addition to Simmel's urban dweller who lives according to the rigors of modern clock time, Benjamin's *flâneur* occupies a privileged role in society. Protesting against the acceleration society, the *flâneur* "is able to break down or see through the surface of the rationalized, commodified experience of bourgeois life" (Dodd and Wajcman, 2016, p. 20). Contrary to Simmel's modern type, the *flâneur* has a more indifferent relationship with time, living in a condition of timelessness.

Finally, Marxist scholars illuminate the relationship between acceleration and the city by emphasising *urban processes*. That is, how expressions of power, politics and economics are being enacted in the context of the city. Critically examining the double process of urbanisation and industrialisation, scholars such Manuel Castells (1977), Henri Lefebvre (1991) or David Harvey (1989) conceptualise urban form and inequality as outcomes of the capitalist mode of production and the distribution of urban resources. Particularly relevant in this context is the notion of "the right to the city", coined by Lefebvre (1996) and revived by Harvey (2008). That is, the city conceived of as an imaginative space from which to envision and move toward a more equitable, collective and participatory process of urbanization. Reimagining the city as a place *for the people* requires recognising shaping forces such as modern capitalism. Most importantly, it requires acknowledging the intimate alliance between spatiality and sociality in creating just and equitable urban societies. Whilst helpful as a contextual layer and an ideal, Metzger (2016) rightly asserts that it has been "proven notoriously difficult to concretize or translate into practice" (p. 585). One of the aims of this book is to unpack the notion of the "right to the city" by shedding light on what "the city" is, which rights are claimed and whose rights are taken into consideration by whom and when.

These classic sociological accounts offer highly relevant analytical frameworks with regard to conceptualising society; that is, *urban* transformations. The 21st century is now shaped by a new pair of forces. Instead of urbanisation and industrialization, Western cities are confronted with the twin forces of urbanisation and datafication. More than ever, the notion of time, in general, and the impact of acceleration, in particular, remains highly relevant. Section 2.2.2. will address the relationship between smart urbanism and temporality. Before engaging with the impact of these aspects on the urban condition, it is critical to acknowledge another "formative context", the transition from postwar Fordist-Keynesianism to the highly entrepreneurial and financialised urbanism, which emerged and expanded from the late 1970s onwards (Morozov and Bria, 2018).

The neoliberal city

Urban planning is based on ideas about society. Tracing and understanding those ideas and imaginaries is central when trying to understand how urban planning agents develop expectations about the future. Imagined futures never emerge in a vacuum but are embedded in socio-economic contexts. According to existing literature in urban sociology, urban planning and political economy, the most formative socio-economic context guiding urban transformations in the last four decades – at least in North America and Western Europe – is "neoliberalism" (Harvey, 1989; Jessop, 2002; Peck and Brenner, 2009; Sager, 2011; Morozov and Bria, 2018). Neoliberalism is generally referred to as "the prevailing pattern of market-oriented, market-disciplinary regulatory restructuring" which "is being realized across an uneven institutional landscape and in the context of heterogeneous, coevolving political-economic processes" (Peck and Brenner, 2009, p. 51).

The transition from centralised and state-led government to decentralised and market-led governance appears to have two immediate consequences for cities. First, in a political climate marked by austerity, fiscal consolidation and a general reluctance to countenance additional direct state borrowing, cities and city regions are increasingly compelled to explore and adopt "financialised" practices in order to attract new investment (Weber, 2010; Guironnet and Halbert, 2014; Snieska and Zykiene, 2015). A second consequence hints at the increasing "entrepreneurialisation" of public administrations. This urban entrepreneurialism (Jessop, 2002) is marked by the "adoption of pro-growth policies and new institutional structures of urban governance, expecting local officials to be enterprising, risk-taking, inventive and profit motivated" (Sager, 2011, p. 154).

The need and pressure for cities to compete and act as "growth machines" (Molotch, 1976) in a stagnating national economy is also reflected in literature examining subnational economic development policy in the UK and Scotland (Loftman and Nevin, 1996; Harding and Le Galès, 1997; O'Brien and Pike, 2015; Waite *et al.*, 2018). Particularly salient is the impact of devolution and the subsequent enhancement of inter-city competition, manifested in the emergence of city region and growth deals since 2011. Comparing 28 city deals agreed between the UK and Scottish governments and city-regional groupings, O'Brien and Pike (2015) argue that "regional and urban policy is being recast as asymmetrical and transactional 'deal-making' and negotiation between central national and local actors unequally endowed with information and resources" (p. 15). As a result of this "deal-making approach" to urban policy, cities and city regions are competing directly with each other for limited public and private capital investment.

Networked urban governance

Urban planning agents such as political leaders, consultants, higher education officials, and tech entrepreneurs are "imagining beings" – architects of the imagination. They lay out the imaginative framework for citizens. They shape the physical and socio-cultural context of expectation formation by engaging in the "politics of urban imagination" (Lindner and Meissner, 2019). The rise of decentralised and more collaborative styles of governance – most clearly exemplified by entrepreneurial and strategic modes of urban development – results in an increasingly complex governance and accountability landscape. Karvonen (2020) points out that "these new modes of networked governance have the potential to promote more democratic modes of urban decision-making but can also obscure responsibility and accountability through diffused forms of political agency and convoluted decision-making processes" (p. 419). The empirical case study underlying this book is no exception. As a result, it is crucial to briefly review the literature that attempts to conceptualise networked urban governance.

One strand of literature highly relevant in this context, examines the relationship between city governance and strategy. Central to debates in this body of literature is an understanding of city administration as a shift from "city government" to "urban governance" (Healey and González, 2005; Stoker, 2011; Harrison and Hoyler, 2014). According to Rhodes (1997) "governance refers to self-organizing, interorganisational networks characterized by interdependence, resource exchange, rules of the game and significant autonomy from the state" (p. 15). The shift from city government to urban governance entails important challenges such as governance gaps, uncertain configurations and a limited capacity to act (Brandtner *et al.*, 2017). Contemporary urban governance appears to be characterised by the need to establish consensus among different urban stakeholders, comprising a "new smart city epistemic community" (Kitchin *et al.*, 2017), by the ability of governance configurations to frame and impact individual governance practices, and by the processes through which governance can be legitimately shaped.

One means to adjust to this changing governance environment is reflected in the strategy-as-practice literature, which advocates the role of "urban strategy". The emergence of urban strategy is closely linked to critiques of traditional urban planning. Famous urban planners such as Le Corbusier, Wright and Howard attempted to establish planning as a field of expertise that is separate from the complexities of daily politics (Hall, 2002). This approach was heavily criticised for its unwillingness to understand and problematise the complex relations between power, rationality and planning (Kornberger, 2012). Contrary to planning, urban strategy constitutes "a body of knowledge concerned with the production of truth, while simultaneously being a political mechanism to mobilize people" (ibid., p. 91). Examining governance through the lens of strategy, Brandtner *et al.* (2017), for instance, argue that texts (such as strategy documents, plans or charts) serve as "discursive devices through which the city administration describes a desirable future, arranges people, objects and topics in time and space and prescribes action for the cast" (p. 1076).

Finally, a third body of literature at the intersection of STS and urban studies, conceptualises cities as constantly emerging, sociotechnical assemblages that are simultaneously discursive, material, temporal, spatial and infused with power dynamics (Latour and Hermant, 1998; Amin and Thrift, 2002; Guy and Karvonen, 2011; McFarlane, 2011; Blok and Farias, 2016; Farías and Blok, 2017). Challenging existing ontological assumptions about "the urban", these authors reject studying the city as an instance of something else (e.g., capitalism or neoliberalism). The notion of assemblage urbanism is also heavily criticised by urban scholars. Brenner *et al.* (2011), for instance, assert that assemblage theorists reject structure in favour of "naïve objectivism" and silence asymmetries, inequalities and hierarchies. Assemblage theorists such as Farías (2011) respond by foregrounding that the adoption of an symmetrical ontology "involves unveiling the actual practices, processes, sociomaterial orderings, reproducing asymmetries in the distribution of resources, of power and of agency capacities" (p. 370), and therefore allows for a radical rethinking – a "democratization of democracy" (ibid., p. 371) – of political practices.

Particularly relevant in the context of this study is Blok's and Farías' (2016) concept of "urban cosmopolitics". Building upon Stengers (2005) and Latour (2004), the authors define urban cosmopolitics as the "search for new urban common worlds of human and nonhuman cohabitation" (Blok and Farias, 2016, pp. 228–229). Contrary to the idea of cosmopolitanism, cosmopolitics does not take the world as given but focuses on "the multiple forces and assemblages that constitute urban common worlds, and on the conflicts and compromises that arise among different ways of composing their forms and limits" (ibid., 2016, p. 2). The cosmopolitical approach to urban governance is highly relevant for the forthcoming analysis since it considers a variety of urban agencies – human and nonhuman – and emphasises the emerging, fragile and constantly shifting nature of urban development. I will return to the conceptualisation of the city as a "multiplicity of urban assemblages" again in Chapter3 when elaborating in more detail on the ontological and epistemological assumptions underlying this study.

2.2.2 Datafication and the city

Late capitalist economies are increasingly shaped by the advent of data-driven, information-led "smart" cities. Similar to the process of industrialisation in the 19th and 20th century, datafication (Flyverbom, Deibert and Matten, 2019) holds the potential to revolutionise the way people live, shape the way people think about society and most importantly, how they imagine futures. The emergence of terms such as "data capitalism" (Myers West, 2019), "platform capitalism" (Srnicek, 2016), "informational capitalism" (Castells, 2010), "surveillance capitalism" (Zuboff, 2015) or "innovation complex" (Zukin, 2020a), are attempts to periodise and localise recent developments in technology and business practices while noting continuity with the capitalist profit motive. In this mode of development, "the source of productivity lies in the technology of knowledge generation, information processing, and symbol communication" (Castells, 2010, p. 17). The digital economy not only legitimises contemporary capitalism more broadly but also fundamentally reshapes economic behaviour on both individual and organisational levels (Srnicek, 2016).

Cities and city regions become central as organisational contexts for the unfolding of data-driven capitalism, but also as active drivers of organisational action in their own right. The fusion of urbanisation and datafication is epitomised in the notion of the "smart" city, a vision for stimulating and supporting innovation and economic growth. Similar to other catch-all terms, the concept of the "smart city" is marked by its nebulous nature and is not reducible to a single meaning. As a result, this book adopts a tentative understanding of smart urbanism by treating it as a broad indicator for the increasing embedding of networked digital instrumentation into the urban fabric. There is a whole body of literature in human geography and urban studies that explores the increasingly networked nature of cities (Mitchell, 1995; Graham and Marvin, 2001; Kitchin and Dodge, 2011; Townsend, 2013). In his decade-long project, former MIT architecture dean William J. Mitchell posits that the digital revolution fundamentally alters patterns of human settlement and experiences of urban life (Mitchell, 1995, 1999, 2003). While presenting invaluable insights on the links between software and city planning and the electronically-mediated self, Mitchell avoids issues of politics such as digital divides. Building on these initial accounts, Graham and Marvin (Graham and Marvin, 2001; Graham, 2016) provide a more in-depth picture of the networked city by shedding light on the changing relations between contemporary cities, infrastructure networks and technological mobilities. Characterising cities as "socio-technical processes", they move beyond classical sociological categories such as class, gender and ethnicity and investigate the often invisible politics of materiality.

Kitchin and Dodge (2011) add to the debate on networked urbanism by making software the central object of analysis. Rather than focussing on the technologies that are enabled by software, the authors argue that software – lines and routines of code – modulates how space comes into being. Doing so, they introduce the concept of code/space to characterise situations when "software and the spatiality of everyday life become mutually constituted, that is, produced through one another" (Kitchin and Dodge, 2011, p. 16). Examples include spaces such as check-in areas in airports or supermarket checkouts. Crucially, their work feeds into the emerging interdisciplinary field of software studies (Manovich, 2000; Fuller, 2003; Lessig, 2006), which seeks to explore and demonstrate the conceptual nature and productive capacity of software and its work in the world from a critical, social scientific and cultural perspective.

The development of micro-electronics-based digital communication networks and information systems – embodied through the smart urbanism agenda – has important spatial and temporal implications. Examining the emergence of the "informational city", Manuel Castells (2010) argues that the material foundations of society are increasingly being organised around the "space of flows" and "timeless time". The predominantly electronic space of flows denotes the "material arrangements that allow for simultaneity of social practices without territorial contiguity" (Castells, 1999, p. 295). Importantly, flows of capital, information and technology induce "timeless time" – the dominant social time of the network society – into social action. The latter "occurs when the characteristics of a given context [...] induce systemic perturbation in the sequential order of phenomena performed in that context" (Castells, 2010, p. 494). Central to this newly emerging temporal and spatial configuration, characterising the informational city, is the notion of big data. Big data bears the hope of a transformation in the understanding and governance of cities "through the creation of a data deluge that seeks to provide much more sophisticated, wider-scale, finer-grained, real-time understanding and control of urbanity" (Kitchin, 2014b, p. 3). In what follows, I will engage with literature examining the smart city agenda, then elaborate on the relationship between data-driven cities and temporality.

The smart city agenda

The "smart city" – one of the most prominent concepts of "the smart" – appears to have successfully captured the public's imagination in the last decade, informing and shaping the work of governmental, civic and entrepreneurial bodies. The smart city agenda builds upon a longer history of computationally networked urbanism, which emerged in the 1970s (Coletta *et al.*, 2018) and in 1992 first appeared as an object of scientific enquiry in the book *The Technopolis Phenomenon: Smart Cities, Fast Systems, Global Networks* (Mora, Bolici and Deakin, 2017; Batty, 2018).

Broadly speaking, there exist two overarching approaches in conceptualising smart cities: the technology-driven or techno-centric approach, and the human-driven or holistic approach (Kummitha and Crutzen, 2017; Mora, Bolici and Deakin, 2017; Kitchin, 2019). Advocates of the former adopt a "technology-first" view, often uncritically pointing out that the deployment of information and communication technology into each activity of the city will improve standards of life (Caragliu, del Bo and Nijkamp, 2011; Zanella *et al.*, 2014; Calzada and Cobo, 2015). Examples of technology-driven urban development projects include Neom[2] in Saudi Arabia, Masdar[3] in the United Arabic Republic or Sidewalk Labs' – a sister company of Google – urban development project in Toronto,[4] which was cancelled in May 2020.

Proponents of the human-driven approach, by contrast, perceive smart cities more critically, emphasising the importance of taking into account issues of power, capital, equality, participation, citizenship, labour, surveillance and alternative forms of urbanism (Hollands, 2008; Kourtit, Nijkamp and Arribas, 2012; Greenfield, 2013; Neirotti *et al.*, 2014; Angelidou, 2015; Datta, 2015; Luque-Ayala and Marvin, 2015; Morozov and Bria, 2018). The most prominent, contemporary example of a human-centric approach to smart city development is in Barcelona.[5]

2 For more information, see: https://www.neom.com/en-us/
3 For more information, see: https://masdarcity.ae
4 For more information, see: https://www.sidewalktoronto.ca
5 For more information, see: https://ajuntament.barcelona.cat/digital/en

Acknowledging the divide in the existing literature on smart cities, Kitchin (2019) criticises proponents of the technology-driven approach for their lack of critical reflection on how smart cities can reproduce particular forms of political economy. At the same time, Kitchin points out that advocates of the human-centric approach "provide little constructive and pragmatic (technical, practical, policy, legal) feedback that would address their concerns and provide an alternative vision of what a smart city might be" (2019, p. 3). Seeking to address this gap, Kitchin (2019) agrees with Luque-Ayala and Marvin (2015) that there is "an urgent need to critically engage with why, how, for whom and with what consequences smart urbanism is emerging in different urban contexts" (p. 2105). Two aspects of the analysis are particularly relevant in the context of this book: first, higher education can act as a powerful urban stakeholder and, second, the dominant epistemological approach to cities needs to be updated.

Concerning the first aspect, scholars acknowledge and critically examine the role of higher education in post-industrial urban redevelopment (Leydesdorff and Etzkowitz, 1995; Etzkowitz, 2002; Katz and Wagner, 2014; Karvonen *et al.*, 2018). Emphasising the role of higher education in the co-production of smart cities, Karvonen *et al.* (2018) stress that university campuses serve as testbeds for smart urban innovation. This is mainly because they "(1) comprise a large, single-owner estate; (2) include a collection of buildings and infrastructure networks that are managed in-house; (3) feed into applied research and teaching activities; and (4) enhance the institutional reputation of the university" (ibid., p. 104). The involvement of the University of Edinburgh in the CRD confirms this observation. The university is the biggest landowner in the city and exerts considerable influence through its five innovation hubs, which constitute the backbone of the DDI Programme. It follows that the future of the city and the university are closely intertwined through the pursuit of transforming the city region into the DCoE.[6]

As for the second aspect, Kitchin (2016) emphasises that current smart city research is often based on a realist epistemology and instrumental rationality. In other words, he points out that data is usually quantitative and one-dimensional in nature and the scientific approach adopted is reductionist and deterministic. As a result, he urges existing knowledge production to "reframe its epistemology to openly acknowledge its situatedness, positionality, contingencies, assumptions and shortcomings, and on the other hand, to comple-

6 The role of scientific expertise in urban development will be dealt with in more detail in Section 3 of this literature review.

ment such knowledge with forms of knowing, such as phronesis and metis" (Kitchin, 2018, p. 223).

Smart urbanism and temporality

Exploring projective agency (Emirbayer and Mische, 1998) in an urban planning context and seeking to understand how datafication affects the process of imagining urban-regional futures, requires taking the relationship between smart urbanism and temporality seriously. Tracing the emergence of the urban tech economy in New York City, Zukin (2020c) argues that during the 2010s "metropolitan cities around the world emerged as dynamic centers of a rapidly growing tech economy" (p. 2). She points out that city officials – political leaders, higher education officials, tech entrepreneurs – form part of an "innovation complex" and imagine a "landscape of innovation" where digital technology leads to innovative research, more efficient public services, the transformation of businesses and, most importantly, the creation of new jobs (Zukin, 2020c). The significance of imaginaries of innovation is confirmed by Söderström, Paasche and Klauser (2014) who emphasise that urban decision-makers are involved in storytelling about smart cities which "shapes the imaginaries and practices of a myriad of actors concretely building the city through particular case studies or pilot projects, decisions and everyday action" (p. 307).

Existing literature in urban studies, sociology, STS and political economy provides highly valuable insights with regard to the temporal analysis of the smart city. Meta-level analyses – as provided by authors such as Castells (2010), Degen (2018) and Kitchin (2019) – shed light on the multiple temporalities that underpin smart urban development. Castells' (2010) notion of "timeless time" represents a helpful starting point for conceptualising the temporal foundation of data-driven cities. According to Castells (2010), the "space of flows" – the predominantly electronic and global materialisation of space – "dissolves time by disordering the sequence of events and making them simultaneous, thus installing society in eternal ephemerality" (p. 497). The notion of "timeless time" resembles Kitchin's (2019) conceptualisation of smart cities as "space-time machines". He points out that "the temporality of the smart city is multiple, heterogeneous, and dynamic, with numerous temporal relations and rhythms unfolding through a diverse set of contingent and relational processes that are intimately enmeshed with spatiality" (Kitchin, 2019, p. 12).

Particularly interesting are his remarks on how the smart city movement shapes urban futures. This occurs in an experimental fashion by prototyping and trialling new technologies, thereby producing "latent futures" (Adam and Groves, 2007). Importantly, Kitchin (2019) emphasises the need for analyses of the politics of time: Whose temporal practices are enhanced by the deployment of smart city technologies? This book engages with the politics of time by critically examining the distribution of projective agency among various urban-regional stakeholders and illuminating how technologies, such as Edinburgh's regional sensor network[7], help mitigate or simply reproduce existing inequalities.

A second group of scholars focuses explicitly on the role of narratives and imaginaries in constructing and managing smart city futures. Existing accounts examine corporate-driven narratives (Klauser, Paasche and Söderström, 2014; Wiig, 2015; Sadowski and Bendor, 2018) and alternative smart city visions (Pasquale and Sadowski, 2015; McFarlane and Söderström, 2017; Sadowski, 2020). Sadowski's and Bendor's (2018) contribution is particularly relevant in the context of this research project. Examining the construction of smart urbanism by the technology companies IBM and Cisco over the last decade, the authors conceptualise the smart city as an "urban sociotechnical imaginary" (Jasanoff, 2015b): "a set of ideas, beliefs, and visions about the future of urbanity" (Sadowski and Bendor, 2018, p. 3). Here, narratives occupy a central role since they entrench the STI of the smart city within "existing cultural norms and moral values, social structures and material infrastructure, political institutions and economic systems, and hopes and aspirations" (ibid., p. 5). Characterising smart urbanism as a "collectively held and performed vision of a desirable future" Jasanoff (2015b, p. 19) hints at the processual nature of the smart city vision. The latter appears to act as an overarching structure, an "umbrella imaginary", which allows for the subsumption of various micro imaginaries and narratives.

2.2.3 Shortcomings and conceptual starting points

Having revised bodies of literature at the intersection of capitalism, the city and datafication, five shortcomings and conceptual starting points emerge. First, it becomes clear that "neoliberalism", during the last four decades, constitutes one of the key explanatory frameworks for examining and conceptualising urban transformations. While taking scholars' use of the term seriously,

7 For more information, see: https://ddi.ac.uk/projects/internet-of-things-research-innovation-service/

it is important to critically acknowledge its conceptual and analytical vagueness. A single case does not seem to exist that cannot be subsumed under the umbrella of neoliberal urbanisation. Venugopal (2015) aptly points out that "if neoliberalism is indeed everywhere and in everything, then it can be productively deployed only as a contextual wallpaper" (p. 169). Consequently, this study recognises the notion of neoliberalism as a "broad indicator of the historical turn in macro political economy" (ibid., p. 181) but refrains from adopting it as a key analytical device.

Second, literature examining the relationship between urban governance and urban strategy provides highly relevant starting points for the forthcoming analysis. Firstly, it emphasises the importance of studying city visions, in general, and urban redevelopment projects, in particular, and sheds light on how urban governance configurations manage and construct uncertain futures. Secondly, this body of literature examines strategy as a practice. It provides analyses of the micro and everyday activities that constitute the labour of strategising, thereby studying strategy meetings, workshops and conversations amongst others. Particularly relevant in this context is the relationship between strategy and performativity. Analysing strategy as a performative practice, Kornberger and Clegg (2011) point out that "strategy turns the arrow of time; the future becomes the condition of the possibility for action in the present" (p. 138). One insight that is especially relevant for this book concerns strategising as a tool, which allows city administrations to move beyond territorial, temporal and jurisdictional boundaries, thereby framing the spaces of intervention and knowledge production in a specific way. The Edinburgh CRD, which envisions the region becoming the DCoE constitutes a case in point. This study adds to existing accounts, mentioned above, by critically examining the uncertain nature of futures. Put differently, strategy tends to be concerned with the politics of what is doable rather than the open space of what can be imagined. Challenging this conservative element, this book, by highlighting imagination and its emancipatory potential and shedding light on how different actors shape the process, can contribute to a more fine-grained understanding of urban strategising.

Third, existing literature dealing with the smart city agenda seems dominated by the artificial divide between a technology-centric and human-centric approach. This research project contributes to the smart city discourse in three ways. Firstly, it provides an empirical account of smart urbanism in the Edinburgh city region – an area that has *not* been studied yet (Karvonen, Cugurullo and Caprotti, 2019). Secondly, it points to the important role of higher education in driving urban-regional economic development. Much of the existing literature has focused on tech corporations as central agents in smart city initiatives (Vanolo, 2014; Wiig, 2015; Sadowski and Bendor, 2018). This study sheds light on the University of Edinburgh as a major force behind urban-regional economic development. Thirdly, it makes a methodological contribution. Complementing existing big data analysis, I deliberately adopt a qualitative approach, based on in-depth interviews, document analysis and ethnographic fieldwork. This will be done by emphasising the role of projective agency or future orientation within urban planning as a promising lens to "know the city". Doing so, this book seeks to account for the multiple, complex and interdependent nature of cities.

Fourth, seeking to develop a better understanding of the organisational context, in which urban planning agents are moving, it is crucial to pay close attention to the relationship between temporality and data-driven cities. The forthcoming investigation contributes to a better understanding of the timescape (Adam, 1998) of smart cities by examining the different temporal relations and rhythms of imagining processes. Furthermore, it is highly interesting to analyse how smart city technologies work with and reconfigure the relationship between the past, the present and the future. Kitchin (2019) notes that "smart cities seek to leverage information about the past and those generated in real time to more efficiently and effectively manage the present and to anticipate and shape the future" (p. 8). Applying Castells' (2010) and Kitchin's (2019) remarks to urban governance, it becomes clear that key challenges entail ensuring connectivity between the electronic space of flows and the physical space of place, but also harmonising and synchronising the temporal relations and rhythms of a diverse set of organisational practices and processes.

Finally, narratives and imaginaries provide promising starting points for the analysis of urban economic development. This study contributes to research at the intersection of STS and urban studies by interpreting cities as complex sociotechnical assemblages and conceptualising urban innovation through Jasanoff's and Kim's (2015) notion of STIs. Contrary to existing accounts, which tend to focus either on the state or on corporate actors as central sites for the evolution and expansion of STIs, this book emphasises the role of cities, more generally, and higher education, in particular, as important drivers of organisational action in their own right when it comes to actualising smart city visions.

2.3 The relationship between scientific expertise and democracy

In this research project, I study the Edinburgh City Region Deal (CRD), a *higher education*-driven smart urban development project, which seeks to transform the city region into the DCoE. Specifically, I study the production of scientific knowledge in urban-regional economic development. The CRD brings into contact two systems – science and politics – with traditionally different functional logics – "learning" and "deciding" – in the common goal to shape and govern this city region's economic futures. Engaging with the decision-makers responsible for crafting data-driven economic futures – principally higher education officials and political leaders – it becomes clear that the distinction between "learning" and "deciding", "knowledge" and "norms" is much more complex than traditional accounts (e.g., Merton, 1973) suggest.

Calling for a close examination of the relationship between scientific expertise and urban-regional politics, this last part of the literature review is guided by the following questions: How does scientific expertise shape urban-regional politics? What role does higher education play in driving datafied imaginaries of "the urban"? How do key decision-makers mediate the science-policy intersection? Tackling these questions, I am going to situate this study within the literature of the sociology of scientific knowledge, the sociology of higher education and political economy, that examines the relationship between scientific expertise and democracy.

2.3.1 Science and democracy

There is widespread consensus among social scientists that modern Western democracies are governed by knowledge and expertise (Bell, 1973; Beck, 1992; Knorr-Cetina, 1999; Jasanoff, 2004, 2017; Grundmann, 2017; Eyal, 2019). Concepts such as "knowledge society" (Bell, 1973) or "risk society" (Beck, 1992) are testament to the deep embedding of scientific expertise, evidence and methods in the very fabric of politics. Ulrich Beck's (1992) "risk society" provides a particularly useful reference point for a better understanding of the relationship between power and knowledge. Seeking to periodise and localise developments in science and technology, Beck argues that post-industrial societies are characterised by a *reflexive* mode of modernisation. Modernisation becomes reflexive during the 21st century. Western societies are no longer concerned with freeing humankind from traditional constraints but essentially have to deal with the risks resulting from techno-economic development itself. Beck

(1992) points out that "in the risk society the unknown and unintended consequences come to be a dominant force in history and society" (p. 22).

Three implications of Beck's analysis are particularly relevant for the underlying study. First, the centre of risk consciousness shifts from the present to the future. Risks such as climate change or global pandemics – most clearly exemplified by the Covid-19 pandemic – lead decision-makers to become active in the present in order to prevent the problems and crises of tomorrow. This is also reflected in urban-regional development. Rather than relying on past knowledge, urban planning agents seek to alleviate and take precautions against the consequences of population growth and climate change by creating sustainable urban-regional visions. Second, there is a "decentring of politics" meaning that "the potential for structuring society migrates from the political system into the sub-political system of scientific, technological and economic modernisation" (Beck, 1992, p. 186). The collaborative style of urban-regional partnerships and the increasingly important role of (supposedly) *non*-political actors, such as higher education, point to the precarious reversal of politics and non-politics.

Finally, reflexive modernisation leads to a demystification of scientific rationality. Beck notes that "science is one of the causes, the medium of definition and the source of solutions to risks" (1992, p. 155). The contradictory nature of techno-scientific development can be illustrated by contemporary urban planning efforts. Seeking to fight climate change and digital poverty, scientific knowledge contributes to the creation and implementation of technologies such as urban-regional sensor networks. At the same time, these technologies produce novel risks around privacy and ethics which call into question science's monopoly on knowledge. It follows that one result of the reflexivity of techno-scientific development is the "demonopolisation of scientific knowledge claims" (Beck, 1992, p. 156). Beck's elaborations on the risk society, particularly his distinction between simple and reflexive modernity, will be picked up in Chapter 7 when I will introduce the notion of "dataist modernity" to characterise the contemporary relationship between technoscience and planning. Building on two historical cases – Le Corbusier's high modernism from the 1920s and Salvador Allende's cybernetic socialism from the 1970s – "dataist modernity" is defined as a form of modernisation which is rooted in an ontology of knowability and which overestimates the power of data in solving complex social problems.

The increasingly fragile role of science in society is picked up by Gil Eyal (2019), who – examining the political uses of scientific knowledge in Western societies – speaks of a "crisis of expertise". This crisis refers to an "unprecedented reliance on science and expertise coupled with increased skepticism and dismissal of scientific findings and expert opinion" (p. 4). The ambivalent nature of expertise points to the social embeddedness of scientific knowledge (Kuhn, 1962), and raises important questions around expert legitimacy and political accountability. This becomes all the more relevant when studying a higher-education driven urban redevelopment project such as the CRD: What kind of expertise do higher education officials exert and how are they held accountable? What role does the public play in crafting economic futures for the Edinburgh city region? The tensions emerging at the intersection of science and politics can be further illustrated by Collins' and Evans' (2017) "wave theory". According to the authors, three different periods of science studies exist. In wave 1, scientific expertise is taken for granted and considered as the pre-eminent form of knowledge production. In wave 2, scientific expertise is acknowledged to be socially constructed. As a result, it calls for a democratisation of knowledge creation. Wave 3 argues that scientific expertise – understood as specialist knowledge – exists. Consequently, Wave 3 proponents seeks to partially reinstate the demarcation between science and politics.

Particularly important for a better understanding of the emergence of smart urbanism in the Edinburgh city region is Collins' and Evans' (2007) discussion of the tensions between the "problem of legitimacy" and the "problem of extension". In other words, how can city leaders ensure citizen participation in knowledge production, on the one hand, while ensuring effective decision-making and consensus, on the other? Building on Collins' and Evans' work, Eyal (2019) identifies four legitimation strategies to fight the "crisis of expertise": 1) exclusion, 2) inclusion, 3) mechanical objectivity, and 4) outsourcing strategies. Legitimacy can be produced by engaging in "boundary work" (Gieryn, 1983); that is, by establishing a clear boundary between science and non-science, experts and laypeople, facts and opinions. Inclusion strategies, on the other hand, seek to render decision-making processes more accountable and transparent by inviting members of the public to participate in the deliberations of expert advisory bodies. The strategy of "mechanical objectivity" is designed to replace the element of human judgment and error as much as possible by use of quantitative measures, standardised tests and objective procedures. Finally, "outsourcing strategies" represent a reaction to the perceived shortcomings of the previous three legitimation strategies. One example of this last approach represents the Intergovernmental Panel on Climate Change (IPCC), which seeks to generate legitimacy by assembling diverse and complementary sources of expertise.

One goal of this book is to define the nature and limits of scientific expertise in urban-regional economic development projects. To this end, I will critically examine how key decision-makers navigate the tensions between the "problem of legitimacy" and the "problem of extension", and what kind of expertise they offer. Having provided a meta-level introduction of the link between science and democracy, I will now zoom in on two themes that are highly relevant in the context of this study: 1) the intersection of the politics and science of the future, and 2) the relationship between science and the city.

2.3.2 The science and politics of the future

Meta processes such as climate change and digitalisation – especially promises around transformative technologies such as AI and robotics – lead to a redefinition of the relationship between science and politics. The increasing demand for policy-relevant work and the scientific supply of future-oriented knowledge demonstrates that the politics and science of the future are intertwined (Wenger, Jasper and Dunn Cavelty, 2020). Urban-regional innovation projects such as the CRD constitute particularly intriguing sites for the study of future-making at the intersection of science and politics. As said above, the promises around data-driven urbanism bring together two systems – science and politics – with traditionally different logics – "learning" and "deciding" (Maasen and Weingart, 2005). Higher education officials, both scientists and full-time managers, interact with political leaders and civil society representatives in their common goal to establish the city region as the DCoE. Scientific knowledge and expertise, supposedly characterised by objectivity and transparency, come into close contact with norms, values and interests.

Jasanoff (2020b) states that the politics of future-making is increasingly shaped by science and technology. Elaborating on the rise of "predictive politics", she emphasises that "collective political imaginations today cannot be separated from the futures promised by science and technology" (ibid., p. 32). This "co-productionist" understanding of future-making is very helpful since it allows for the conceptualisation of the DCoE vision as an "urban-regional sociotechnical imaginary" (Jasanoff, 2015c, p. 4). Adopting Jasanoff's (2004) "co-production framework", this study sheds light on higher education as a central site for the evolution and expansion of STIs. In what follows, I will engage with key contributions in STS, sociology and political theory, that problematise the science-politics intersection by 1) elaborating on the relationship between expertise and legitimacy, 2) shedding light on boundary work, and 3) focusing on the link between innovation and imagination.

Expertise and legitimacy

One strand in STS engages with the question of what constitutes the legitimacy of scientific experts (Weingart, 1999; Hoppe, 2005; Jasanoff, 2005, 2011; Collins and Evans, 2007; Bandola-Gill, 2020). The key message that these accounts convey is that scientific knowledge is never objective or value-free but always embedded in sociopolitical structures and norms. Weingart (1999) highlights that the interaction between science and politics leads to both "the politicisation of science" and "the scientisation of politics". Science becomes guided by political priorities and political decision-making is expected to be based on scientific evidence. Building on Jasanoff (2005), Bandola-Gill (2020) argues that expert legitimacy is constructed in the "symbiotic negotiation between technocratic and political modes of accountability" (p. 14) unfolding on three levels – epistemic, individual actor and institutional. This insight is helpful since it illustrates that expert legitimacy is not simply an individual quality but rather a multifaceted and culturally determined concept. Jasanoff (2003a) goes on to define expertise as

> not merely something that is in the heads and hands of skilled persons, constituted through their deep familiarity with the problem in question, but rather that it is something acquired, and deployed, within particular historical, political, and cultural contexts (p. 393).

This book investigates the nature of scientific expertise in the construction of the DCoE vision. Doing so requires not only asking how expert legitimacy is constructed but also elaborating on what counts as expertise in the first place. Two conceptualisations of expertise are particularly noteworthy in this context. The first one is Collins' and Evans' (2007) distinction between "interactional" and "contributory" expertise. Seeking to provide a more nuanced understanding of expertise, the authors define "interactional expertise" as the "ability to master the language of a specialist domain in the absence of practical competence" (p. 14). "Contributory expertise", in turn, "enables those who have acquired it to contribute to the domain to which the expertise pertains" (p. 24). This two-fold understanding of expertise involves a *realist* and a *relational* element since it, on the one hand, insists that expertise is a real and substantive possession (contributory expertise) but, on the other hand, emphasises that expertise is also acquired by virtue of being member of a specialist group (interactional expertise). Collins' and Evans' work theorises expertise primarily on the basis of experimental scientists (Grundmann, 2017). It is important, however, to recognise that the laboratory as a site of knowledge production and the corresponding research institute are not the only sources of expert knowledge, particularly regarding politico-economic decision-making.

Another body of academic work moves beyond science-based notions of expertise by emphasising the role of "lay expertise" and stakeholder participation (Wynne, 1996; Callon, 1999; Jasanoff, 2003b; Corburn, 2005). Building on Wynne's (1996) seminal study on Cumbrian sheep, Callon (1999) problematises the divide between specialists and non-specialists by emphasising the complementarity of expert and lay expertise. He distinguishes between three models of knowledge production and points out that the supposed crisis of confidence in technoscience is actually a crisis of knowledge production regimes. In other words, contemporary knowledge production regimes either rely on lay people's full trust in scientists (and therefore absence in knowledge production) or lay people merely enhancing or completing scientific knowledge production. The third knowledge production regime, the "co-production of knowledge model", in turn, "tends to overcome these limits by actively involving lay people in the creation of knowledge concerning them" (Callon, 1999, p. 89). Jasanoff (2003b) adds to this line of thought by introducing the notion of "technologies of humility". Complementing predictive technologies such as risk assessment or climate modelling, these *social* technologies "require not only the formal mechanisms of participation but also an intellectual environment in which citizens are encouraged to bring their knowledge and skills to bear on the resolution of common problems" (p. 227). Put differently, "technologies of humility" conceptualise human beings as imaginative and active agents and promote the inclusion of personal experience, place, history and social connectedness as important factors in the production of knowledge.

Mediating the science-policy boundary

The emergence and implementation of the CRD is shaped by the intimate entanglement between science and politics. This book seeks to provide a detailed description and typology of the processes and agents involved in creating, dispersing and stabilising data-driven urban-regional futures in the Edinburgh city region. Doing so, it is important to briefly review academic perspectives in STS, sociology and political science that engage with actors and processes along the science-policy intersection. Gieryn's (1983, 1999) concept of "boundary work" provides a helpful starting point. Emphasising the social construction of science, Gieryn (1983) points out that "its boundaries are drawn and redrawn in flexible, historically changing and sometimes ambiguous ways" (p. 781). Operationalising the science-policy intersection through Gieryn's concept of "boundary work" seems highly useful since it allows us to shed light on how higher education officials demarcate science from non-science, thereby avoiding potential issues of democratic accountability. One key question guiding this book is the following: Where does the role of science end and that of politics or policy begin?

Individuals involved in moving and translating knowledge along the intersection of politics and science are conceptualised as "knowledge brokers" (Meyer, 2010), policy entrepreneurs (Mintrom, 1997; Kingdon, 2014) or "translation agents" (Jasanoff, 2015c). "Knowledge brokers", for instance, are defined as "people or organizations that move knowledge around and create connections between researchers and their various audiences" (Meyer, 2010, p. 118). One analytical framework that seems particularly helpful for theorising the role of higher education officials in urban-regional development is Callon's (1984) "sociology of translation". According to Callon, the process of translation consists of several moments: defining actors by assigning roles to them; specifying and testing these roles; and rendering actors mobile. As a result, a few actors become "obligatory passage points" (Callon, 1984, p. 196) – they "obtain the right to express and to represent the many silent actors of the social and natural worlds they have mobilized" (ibid., p. 224). Callon's (1984) sociology of translation will be picked up again in Chapter 5 when I will focus on the role of four senior higher education officials responsible for creating and dispersing the DCoE vision.

Imagination and innovation

Finally, this study sheds light on the relationship between imagination and innovation. Innovation – *data-driven* or *technological* innovation in particular – has become a central arena for politics and an uncontested value in contemporary Western capitalist societies (Godin and Vinck, 2017; Pfotenhauer and Jasanoff, 2017a; Pfotenhauer and Juhl, 2017; Rammert *et al.*, 2018; Zukin, 2020a). In fact, Pfotenhauer and Jasanoff (2017a) point out that "it has become virtually impossible to talk about economic development or social progress in terms that do not invoke, explicitly or implicitly, the need for innovation" (p. 784).

Tracing the rise of the urban tech economy, Zukin (2020a) asserts that "what cities are envisioning today is nothing less than the urbanization of Silicon Valley, an imaginary based on placing new digital technologies in dense, strategic clusters, creating new cultures of innovation and production, and capturing the economic rewards" (p. 3). And Rammert *et al.* (2018) underscore that "innovation is transcending its traditional boundaries to become the major driving force in the society of the future" (p. 3). Studying the production of a specific urban-regional STI, this book focuses on dynamics of urban-regional innovation and pays particular attention to its limits.

One strand in STS that is highly relevant for this endeavour adopts a co-productionist lens for the analysis of innovation policy; thereby affirming that technoscientific and sociopolitical orders are mutually constitutive and dependent. Examining the implementation of the MIT model of innovation in the UK, Portugal and Singapore, Pfotenhauer and Jasanoff (2017a) identify culturally specific "imaginaries of innovation". That is, "epistemic and political resources for defining a community that shares a common (and hopefully better) sociotechnical future, and that is on its way to attaining it through innovation" (p. 788). The findings of this study are relevant for two reasons. First, the authors treat innovation as a locally constructed concept while recognising innovation models' global reference points. In other words, locality matters – there is not a universal set of innovation mechanisms. Second, the study demonstrates power differentials in "the global political economy of innovation". Put differently, innovation models such as the MIT or Silicon Valley one "act as common reference points for a global community of innovation practitioners and policy makers" (Pfotenhauer and Jasanoff, 2017a, p. 803). Both findings indicate that innovation processes constitute prime sites for the operation of STIs.

Another insight in current STS research on innovation, that is highly relevant for this research project, refers to the inherently political nature of innovation. Examining state-led technology and innovation policy, Pfotenhauer and Juhl (2017) challenge accounts which conceptualise technoeconomic progress as divorced from political intent. Instead, they argue that "innovation is political 'all the way through'" (p. 88), ranging from the "recognition of the collectively binding decisions [humans] make through innovative technologies" to the "inevitable controversies accompanying an inclusive, democratic approach to innovation-based progress" (ibid.). It follows that this project's empirical focus, data-driven urbanism, does not constitute an apolitical process but should in fact be grasped as an "urban political technology" (Pollio, 2016) allowing decision-makers to advance specific socio-economic futures.

2.3.3 Science and the city

In this book, I shed light on the close interconnections between knowledge production and urban-regional administration. The CRD provides a stage for the interplay of two actors: science and the city, higher education and local-regional government, the University of Edinburgh and the Edinburgh city region. Dierig *et al.* (2003) provide a helpful account for the conceptualisation of the relationship between science and the city. The authors discuss four ways in which the latter intersect: 1) the relationship between scientific expertise and urban politics, 2) science's role in the cultural representation of the city, 3) the embedding of scientific activity in urban socio-material infrastructure, and 4) the interaction between science and everyday urban life.

The first intersection – scientific expertise and urban politics – points to the role of scientific experts as regulators and shapers of urban space. Not only is scientific authority involved in urban planning but urban space simultaneously informs scientific knowledge production. Studies in STS that examine the impact of scientific imaginaries in urban development projects include Bruno Latour's (1996) *Aramis or the Love of Technology* or Wakeman's (2003) *Dreaming the New Atlantis*. Second, cities and city regions are also constituted by their representations. Put differently, scientists play a role in creating and shaping representations of the urban. The CRD and its ambition to transform the city region into the DCoE illustrates a case in point for scientists' role in establishing new ways of seeing and knowing the city (region).

The third intersection points to the relationship between places of knowledge and their urban contexts. As mentioned above, the University of Edinburgh constitutes the biggest landowner in the city and is literally *building the future* by expanding its physical presence and visibility in the city region. Finally, everyday city life and scientific knowledge production are closely intertwined. One example that underscores how urban culture can feed into scientific practice is Georg Simmel's (1903) classic essay on *The Metropolis and Mental Life*. Reflecting on these four intersections demonstrates the co-productionist nature of science and the city: science makes the city but the city also makes science. Dierig *et al.* (2003) go as far as saying that "seeing science as an urban practice also means studying the practice of science as part of the urban setting" (p. 4).

One body of literature that allows for a better demarcation of the underlying research problem examines the relationship between higher education and urban-regional economies (Etzkowitz and Leydesdorff, 1995; Berman, 2012; Capello, Olechnicka and Gorzelak, 2012; Audretsch, 2014; Goddard *et al.*, 2014). Three themes covered in this literature are particularly important for the forthcoming analysis: 1) the rise of higher education as driver of economic growth, 2) the multi-scalar nature of universities, and 3) the notion of the "civic university" (Goddard, 2009). Authors such as Berman (2012b) and Audretsch (2014) demonstrate how the role of higher education institutions (HEIs) evolved from being a resource to policy and economy to being an engine of economic growth. Berman (2012b) argues that there are two main reasons why academic science moved toward the market: first, government policies encouraged the commercialisation of scientific practices and, second, policymakers started to identify scientific and technological innovation as key driver of the economy. The UK government's *Research and Development Roadmap* (2020) and its ambition to increase public investment in R&D to £22bn per year by 2024/25 supports the growing importance of innovation as a key driver of economic prosperity. Similarly, Etzkowitz's and Leydesdorff's (1995) "triple helix model" of innovation emphasises new institutional configurations between academia, government and industry, in which universities take up diverse development, innovation and regional leadership functions. This book builds upon this line of argument and contends that HEIs represent dominant urban-regional actors, thereby moving beyond their traditional missions of research and training.

A second theme that is highly significant when examining higher education-driven, urban-regional development projects constitutes the multi-scalar nature of universities. Goddard and Vallance (2013) point out that universities represent "place-based institutions with connections into the different social and institutional spheres of [their] locality" and act as "generative node[s] in national and international flows of knowledge and people (especially highly-mobile students)" (p. 2). In other words, unlike city authorities, *global* HEIs such as Edinburgh University do not operate within bounded territories. Consequently, their spatial interests might not necessarily be local. Scholars such as Addie *et al.* (2015) or Power and Malmberg (2008) problematise this tension by distinguishing between innovation in general and *urban regional* innovation in particular. One goal of this study is to critically examine the role of Edinburgh University in urban-regional innovation. Put differently, what impact does the CRD have on local communities? Do scientific and technological innovations permeate local industries, and thereby positively affect the lives of residents or simply reproduce existing inequalities?

Finally, one theme that has recently attracted attention is the idea of the "civic university" (Goddard, 2009; Goddard, Vallance and Kempton, 2012). Challenging the concept of the "entrepreneurial university", the "civic" element implies the attempted move from the "triple helix" (government, academia, industry) to the "quadruple helix" (government, academia, industry, *civil society*) of innovation. Goddard *et al.* (2012) suggest that the civic university is an organisation that places its research and education in the service of the community by engaging as a whole with its surroundings, partnering up with other local HEIs and actively incorporating locality into its identity. Aspiring to the ideal of the "civic university" can be read as an attempt to deal with the tensions arising from the multi-scalar nature of HEIs. This becomes all the more clear when engaging with urban-regional decision-makers. Depending on context, multiple interpretations and imaginaries of the University of Edinburgh co-exist. It follows that this study will disentangle the multitude of imaginaries with a particular emphasis on separating rhetoric from reality regarding the use of "civic university" imaginary (see Chapter 6).

2.3.4 Shortcomings and conceptual starting points

Existing accounts in STS, sociology and political economy make excellent contributions to a better understanding of the dynamics between scientific expertise and democratic politics. In what follows, I want to elaborate on helpful conceptual starting points and a range of shortcomings which this book seeks to address.

First, this study will shed light on the relationship between scientific expertise and democratic legitimacy. Eyal's (2019) "crisis of expertise" provides an extremely helpful conceptual layer for analysing the nature of expertise and the legitimation strategies involved in mobilising expertise. Building on existing accounts that acknowledge the "politicisation of science" and "the scientisation of politics" (Weingart, 1999), this book aims to shed more light on how expert legitimacy is actively constructed in urban-regional development. In doing so, it is crucial to gain a better understanding of the balance between scientific ideals and policymaking. Jasanoff (1990) indicates that "experts themselves seem at times painfully aware that what they are doing is not 'science' in any ordinary sense, but a hybrid activity that combines elements of scientific evidence and reasoning with large doses of social and political judgment" (p. 229).

In Chapters 5 and 6, I will explore this tension by 1) conceptualising higher education officials as translation agents mediating the science-policy boundary, and 2) focusing on the role of Edinburgh University caught up in competing imaginaries of the 21st century university. In addition, it is important to examine *internal* boundary work. That is, what forms of expertise were and are excluded and who decides what evidence is actually used? How is evidence mobilised in urban-regional future-making? This book will critically unpack the nature of evidence-based policymaking in the Edinburgh city region by emphasising that evidence is *not* value-free but reflects evidence providers' social embedding and normative standpoints.

Second, existing accounts in STS exploring the interplay of science and the city have neglected the role of HEIs as important agents in urban-regional governance. Since its inception as a field in the 1970s, STS considered the laboratory as the central site of knowledge production and technological development. One can surmise that this has led to a neglect of the university as the context in which laboratories themselves – as well as other spaces of knowledge production, such as libraries – often still need to be embedded. Needless to say, universities have been studied extensively, especially by higher education researchers and urban planning scholars, but mostly as formal organizations rather than as concrete, materialized, and highly complex social places with their own logic. This study views HEIs as indispensable spaces for generating research cultures within which science and technology can then be developed, debated and disseminated. In doing so, it builds on existing concepts such as "knowledge-based urban development" (KBUD), which advocate the eminent role of knowledge related to economic, societal, spatial and governance domains in the determination of future development paths for cities (Yigitcanlar, Velibeyoglu and Baum, 2008; Yigitcanlar, 2011; Carrillo *et al.*, 2014).

Furthermore, drawing on Berman (2012b) and Audretsch (2014), who demonstrate how the role of HEIs evolved from being a resource to policy and economy to become an engine of economic growth, this book argues that global HEIs such as the University of Edinburgh move beyond their traditional missions of teaching and research, and actively contribute to urban-regional economic development. In Chapter 4, I will trace the emergence of the DCoE vision and highlight the enormous influence of higher education officials in establishing new ways of seeing and knowing the city. One key question is: How can Edinburgh reinvent itself to embrace imaginaries of data-driven futures when its modus operandi has been its historical past? Furthermore, the multi-scalar nature of Edinburgh University will be addressed by examining how, depending on context, actors interpret it as a local, national or global institution.

Figure 5: Literature review summary | Source: Own illustration

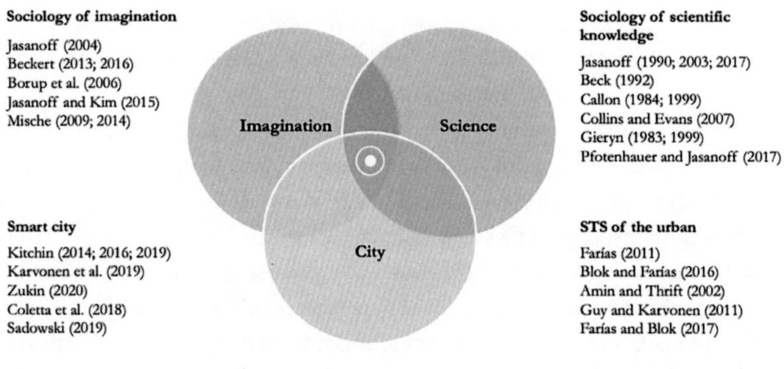

Third, this book conceptualises urban-regional innovation as an inherently political process. In line with critical innovation studies, which challenge "the pro-innovation bias" (Godin and Vinck, 2017), I aim to shed light on resistance to innovation, the unintended consequences of innovation and outright failures of innovation. Adopting a co-productionist lens acknowledges that technoscientific and sociopolitical orders are mutually constitutive and dependent. I will highlight the close interconnections between scientific elites and political power. As a result, I ask: What are the limits of the CRD? What counter-imaginaries existed and still exist for the future of the city region? In addition to that, it is also highly relevant to analyse how local-regional STIs are linked to national imaginaries of innovation. In other words, how does the collective political imagination of Scotland as a (potentially independent) nation relate to the DCoE vision? Do Scots perceive data-driven innovation as paving the way for sustainable economic growth, or is the CRD simply a tool for the University of Edinburgh to present itself as a "Scottish" institution? In summary, it is paramount to acknowledge that the sites and objects of politics are no longer simply those of formal state institutions but also lie in the hands of increasingly important players such as higher education that fall outside the bounds of meaningful political control. Consequently, urban-regional innovation must be conceptualised as a political category and subjected to normative standards such as transparency and accountability.

3. Conclusion

In this chapter, I have suggested that academic literature at the intersection of the sociology of imagination and science (Beck, 1992; Jasanoff, 2004; Beckert, 2013a; Jasanoff and Kim, 2015; Beckert, 2016; Eyal, 2019) has applicability towards a new kind of data-driven urban-regionalism (see Figure 5). To this end, I have reviewed relevant theoretical debates and empirical discussions along three broad thematic strands: 1) capitalism's temporal order and decision-making, 2) data-driven innovation and the city, and 3) the relationship between scientific expertise and democracy. In what follows, I want to briefly summarise the key findings and link them to the following empirical chapters.

Existing academic accounts in sociology, STS, political economy and urban geography make excellent contributions to a better understanding of the interplay between science, imagination and the city. There are at least three aspects which prove extremely helpful for the underlying study. First, academic perspectives within the sociology of imagination underscore the importance of researching futures in action. Scholars such as Jens Beckert and Sheila Jasanoff demonstrate that expectations, narratives and imaginaries provide highly promising units of analysis for a better understanding of the dynamics of capitalist societies. Second, empirical studies in the sociology of scientific knowledge highlight that the politics and science of the future are closely intertwined. Shedding light on the science-policy boundary, these accounts illuminate the complex relationship between scientific expertise and democratic legitimacy. Finally, scholars at the intersection of STS and urban studies conceptualise cities as sociotechnical assemblages, thereby emphasising the distribution of agency among human and non-human actors as well as the constantly emerging quality of "the urban".

Building on the reviewed literature, this book attempts to make several contributions. First, it defines higher education as an important site for the evolution and expansion of urban-regional STIs. This includes tracing the construction of the Edinburgh city region's innovation apparatus (Chapter 4), an investigation of scientific expertise in regional economic development by drawing on Callon's (1984) sociology of translation (Chapter 5), and developing the concept of "institutional eclecticism" to characterise increasingly competing imaginaries of the modern university (Chapter 6). Second, this study intends to contribute towards a better understanding of the role of uncertainty in politico-economic decision-making. This will be done by challenging the adoption of uncertainty as a conceptual starting point for a sociological analysis of innovation economies through illuminating the "politics of (un)*certainty*" in urban-regional governance. Finally, this book contributes to a better understanding of the interplay of technoscience and urban-regional planning by contextualising the emergence of Edinburgh's innovation apparatus with earlier attempts to steer society, and by situating empirical findings within a contemporary moment I call "dataist modernity" (Chapter 7).

Before interrelating conceptual starting points with empirical findings, I will – in the next chapter – paint the picture of my research journey. This includes a more informal and reflexive account, in which I will introduce the underlying empirical case study, elaborate on the study's rationale and challenges involving the data generation and analysis process, and discuss ethical considerations that have shaped the research process.

Chapter 3: Studying Projective Agency in the Edinburgh City Region

1. Introduction

This chapter tells the story of my research journey. Having critically reviewed theoretical debates and empirical discussions along three thematic strands of literature, it is essential to elaborate on the research design of this book. To this end, I will engage with the following questions: What was I trying to find out? Which questions and considerations have structured the research process? Who did I speak to and why? How did my own positionality affect the research process? Finally, what went wrong?

The research problem of this project is concerned with projective agency in an urban planning setting. As introduced in the previous chapter, projective agency denotes "the imaginative generation by actors of possible future trajectories of action, in which received structures of thought and action may be creatively reconfigured in relation to actors' hopes, fears, and desires for the future" (Emirbayer and Mische, 1998, p. 971). Projective agency constitutes the primary unit of analysis and is operationalised by use of Jasanoff's and Kim's (2015) notion of "sociotechnical imaginaries". STIs – likewise mentioned before – are defined as "collectively held, institutionally stabilized, and publicly performed visions of desirable futures, animated by shared understandings of forms of social life and social order attainable through, and supportive of, advances in science and technology" (ibid., p. 4). This book investigates the ways in which urban planning agents imagine futures by examining the construction and diffusion of a specific urban-regional sociotechnical imaginary – establishing the Edinburgh city region as the DCoE.

Given this primary research interest, the research problem at hand required answers that were descriptive and explanatory. Following an interpretive approach I therefore adopted an inductive logic of inquiry. This implies moving from observation to explanation (Blaikie and Priest, 2017). Conceptualising imagination as a social practice, the aim was to, first, build a body of rich data and then, second, construct and present an explanation (Mason, 2002).

This chapter aims to provide a chronological account of the research conducted for this book. To this end, it is divided into three parts. I begin by elaborating on how I developed the research project. This includes a brief section on my personal journey as well as the introduction of the empirical case study. Afterwards, I will discuss the methodological approach underlying this study. To do so, I will explain which methods were adopted and to what end. Sections 2.3–2.5. will discuss the data analysis stage, thereby reflecting on the study's coding strategy and ethical considerations. Finally, concluding remarks follow. Throughout the chapter, I intend to adopt a reflexive and more narrational mode in order to illuminate significant observations, expected and unexpected issues and the role of my own positionality within the research process.

2. Research journey

The research journey can be divided into three parts: 1) project development, 2) data generation, and 3) data analysis. Importantly – and unsurprisingly – the research journey did not unfold in a linear fashion, moving frictionless from point A to point B. Instead, it was characterised by a constant back and forth, shaped by empirical findings and external factors such as the Covid-19 pandemic.

2.1. Project development

My interest in cities, expectations and imagination began during my Master's in International Political Economy at the University of Warwick. Adopting a critical sociological perspective in my master's dissertation, I analysed how people imagine economic futures and what role social acceleration plays in this quest. Focusing on the urban condition, I argued that urban gardening projects such as the "Incredible Edible Todmorden"[1], by creating alternative spaces of reproduction, foster the imagination of alternative economic futures. Building upon my strong interest in the role of cities as places of

1 For more information, see: https://www.incredible-edible-todmorden.co.uk

emancipatory economic imaginations, I developed a PhD project which set out to bring together the sociology of expectations and urban planning. Fascinated by the interplay of temporality and spatiality, I initially aimed to relate two existing explanations for the dynamics of capitalism, Jens Beckert's (2016) "fictional expectations" and Hartmut Rosa's (2013) "social acceleration", and to apply them empirically to the urban environment. I was intrigued to find out how expectations about the future develop *in* the urban landscape, and, at the same time how they condition the imagination for the development *of* urban landscapes. Consequently, I asked: How do fictional expectations develop (in) the urban landscape?

Arriving in Edinburgh in September 2018, I was struck by the social and physical presence of the University of Edinburgh. Whether walking across George Square, spotting the ocean from New College or following James Hutton's footsteps around Arthur's Seat, it became clear that the university makes the city and the city makes the university. I had no difficulties in imagining how Edinburgh developed from a modest city into a hub of government, medicine, finance and education. I had no difficulties in grasping its reinvention from "auld reekie" to enlightened New Town. What was much more difficult to imagine, however, was the idea that Edinburgh would become a digital hub, a centre of innovation, let alone the DCoE. Everyday urban life seemed to be shaped by achievements of the past – embodied and reproduced in the urban landscape via buildings, street names or the nutty scent of the breweries and distilleries. How can a city reinvent itself in order to embrace imaginaries of data-driven futures when its modus operandi has been focused on its historical pasts?

One attempt to provide a convincing response to this question can be found in the university's decision to establish the Edinburgh Futures Institute (EFI). Located in the Old Royal Infirmary of Edinburgh, which is currently being renovated, the EFI constitutes one of six innovation hubs underlying the Edinburgh City Region Deal (CRD) (Figures 6 and 7). In many ways it embodies the past and the future, the old and the new Edinburgh. The EFI depicts the university's attempt to have a different kind of conversation about data-driven innovation (DDI) between those scholars and practitioners who are immersed in high-tech epistemologies and those coming from more critical traditions. Following Geddes' (1904) ideal of "civics", the EFI is supposed to live up to the ideal of the "porous" university, that is, engaging with local residents and placing its research and education in the service of the community.

Figure 6: Image of the Old Royal Infirmary building | Source: Boon Law, 2008, Flickr creative commons

Figure 7: Architect's representation of finished building | Source: Bennetts Associates

Given its prominent location at the heart of Edinburgh's Old Town, right next to the School of Social and Political Science, the EFI provided an unexpected yet incredibly helpful empirical access point. Unaware of the Edinburgh CRD at this point, I started to conduct a range of pilot interviews to gain a better understanding of the role of expectations in urban planning. To this end, I interviewed three representatives from the EFI to examine how they imagine the (uncertain) future. Themes generated in this pilot study include the role of interdisciplinarity as a device for managing uncertainty, the importance of stories and narratives when accessing and shaping futures, and the tension between human- and data-driven futures. Most importantly, I managed to build rapport, and thereby secured not only a potential avenue for further access to relevant key stakeholders but even became part of the EFI film project *Closes and Opens: A History of Edinburgh's Futures.*[2] As a result of this initial pilot study, I became aware of the significance of the Edinburgh CRD and subsequently decided to select it as my empirical case study.

2.1.1 Empirical case study

I now want to introduce the empirical case study underlying this research project. I will briefly describe and situate the case. Afterwards, I want to justify the CRD and its focus on DDI as promising units of observation. This PhD project intends to take projective agency in an urban planning setting as its primary unit of analysis. To that end, it examines the implementation of a contemporary urban redevelopment project, the Edinburgh and South-East Scotland CRD. The CRD represents a £1.3bn investment over the next fifteen years, which is designed to accelerate inclusive growth through the funding of infrastructure, skills and DDI. It consists of five themes (Figure 8) – (1) research, development and innovation, (2) integrated regional employability and skills programme, (3) transport, (4) culture, and (5) housing – of which the first area of focus receives the lion's share (£791 million).

The city region comprises six local authorities: Fife, East Lothian, Midlothian, the Scottish Borders, West Lothian and Edinburgh. The latter includes a population of 1.4 million people (24% of the population of Scotland) and contributes approximately £36bn per year to the Scottish and UK economies. Seeking to provide the jobs of the future and address key economic barriers and issues constraining the region's economic potential, the CRD aims at establishing the city region as the DCoE (The City of Edinburgh Council, 2018).

Unit of observation I: The City Region Deal

Urban planning policy in the UK has been subject to frequent restructuring with regard to governance arrangements and policies across different governments (Martin *et al.*, 2019). Examining subnational economic development policy since the 1970s, Martin *et al.* (2019) point out that "the territorial focus of subnational economic development policy has changed frequently from

2 The movie can be accessed here: https://www.youtube.com/watch?v=ombo3njkfL8

regionalism to localism, back to regionalism, then localism and most recently to city-regionalism" (p. 52).

Since the coalition government in 2010, city region and growth deals became the preferred model of subnational economic development in the UK (Waite *et al.*, 2018). They can be defined as "bespoke packages of funding and decision-making powers negotiated between central government and local authorities and/or Local Enterprise Partnerships and other local bodies (Ward, 2018, p. 4) (Figure 9). As of 2021, there are twelve city region and growth deals (5 signed and 7 in the development stage) in Scotland. UK and Scottish governments, together with councils and regional partners committed funding, worth a total of £5.2 billion (Audit Scotland, 2020). The Glasgow CRD, agreed in August 2014, constituted the first Scottish city region and growth deal. The Edinburgh CRD was officially signed on 7 August 2018 by the First Minister of Scotland, the Prime Minister and city region leaders.

*Figure 8: Edinburgh and South-East Scotland CRD financial summary | Source: The City of Edinburgh Council (2018) *Includes £120m for Sheriffhall roundabout to be delivered by Transport Scotland*

Theme	Government contribution (£m)	Partner contribution (£m)	Total amount (£m)
Research, development and innovation	£350	£441	£791
Integrated Regional Employability and Skills Programme	£25	N/A	£25
Transport	£140*	£16	£156
Culture	£20	£25	£45
Housing	£65	248	£313
Total	£600	£730	£1,330*

The emergence of the Edinburgh CRD must also be understood in light of the UK government's increased focus on innovation as key driver of future economic growth. This is reflected in the *Industrial Strategy 2017* and the *Research and Development Roadmap 2020*. Seeking to close the gap between knowledge generation and knowledge application and to fight regional inequalities, the

UK government committed to increasing investment in R&D to 2.4% of the GDP by 2027 and to public funding for R&D to £22 billion per year by 2024/25 (UK Government, 2020). Innovation-driven CRDs, such as the Edinburgh one, constitute key tools to advance this agenda. Not only do they shed light on the UK's innovation ecosystem and the more active role of the state in shaping economic policy, but they also underscore the importance of higher education institutions (HEIs) such as the University of Edinburgh in shaping future economic growth. A detailed discussion of past and contemporary shifts in UK innovation policy will be provided in the next chapter.

Figure 9: City region and growth deals | Source: Audit Scotland (2020)

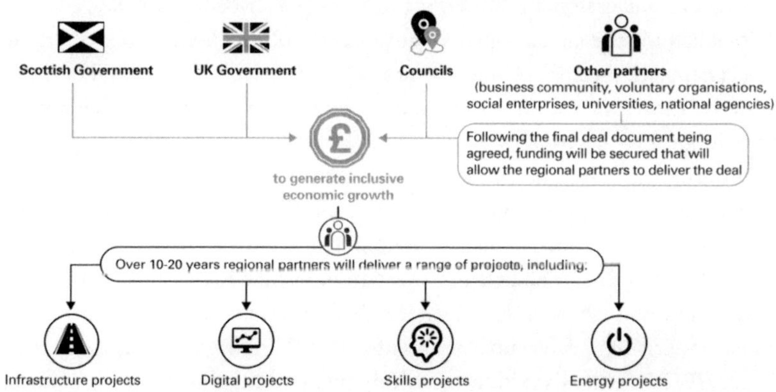

For now, I want to briefly examine the unique characteristics of the CRD by elaborating on identified challenges of the city region as well as proposed solution approaches. From the outset, it becomes clear that the CRD's ambition is first and foremost economic in nature. Early discussions about the CRD arose in 2015, acknowledging the background of a digital economy growing two to three times faster than the economy as a whole. Comparing the Edinburgh CRD with the other eleven city region and growth deals in Scotland, it becomes apparent that this deal sticks out for at least three reasons. First, it has attracted the largest investment. Second, its key focus – contrary to, for instance, Glasgow with a focus on transport – is on DDI. Most importantly, the CRD differentiates itself from other city region and growth deals by the significant level of involvement of higher education, most notably, the University of Edinburgh. It constitutes the first higher education-driven CRD in the UK and acts as a blueprint for other city regional deal proposals. Therefore, it has relevance far beyond the Edinburgh city region.

Unit of observation II: The DDI Programme

Seeking to tackle challenges such as slow growth and skills inequality, DDI is claimed to constitute the key means to ensuring inclusive growth. This is reflected in the DDI Programme[3], which aims to enhance the data capability of the city region across ten key industry sectors through five areas of activity: talent, research, adoption, data and entrepreneurship (TRADE). The DDI Initiative constitutes the largest part of the CRD and receives a total of £661m (Data Driven Innovation Programme, 2022). Designed to span ten years, it is delivered by the University of Edinburgh and Heriot-Watt University. Key DDI activities include the Global Open Finance Centre of Excellence[4], the Wayra AI Blockchain Accelerator[5] or the Advanced Care Research Centre[6]. All of these initiatives constitute multi-disciplinary research programmes, which aim to drive innovation in chosen sectors (e.g., finance, AI, healthcare).

Attempting to transform the city region into the DCoE, six DDI hubs have been created: the Bayes Centre (Figure 10), the EFI, the Easter Bush campus, the Usher Institute for Population Health Sciences Informatics and the National Robotarium. In addition to these more visible materialisations of the DDI Programme, there is also a more hidden, invisible side to the DCoE imaginary. Underlying the five innovation hubs is the Edinburgh International Data Facility (EIDF)[7] – the sixth innovation hub – which aims to provide the required underpinning data capability, computing and data storage infrastructure to understand trends and behaviour across the city region. In many ways, the EIDF's Advanced Computing Facility (Figures 11 and 12) depicts the real centre of power since it promises that academia, industry and government will benefit from the collection, use and analysis of data. The investment into the six innovation hubs – a £270m government capital investment to the University

3 For more information, see: https://ddi.ac.uk

4 For more information, see: https://www.globalopenfinance.com

5 For more information, see: https://www.ed.ac.uk/bayes/our-community/past-collaborations/wayra-uk

6 For more information, see: https://www.ed.ac.uk/usher/advanced-care-research-centre

7 For more information, see: https://www.epcc.ed.ac.uk/research/data/eidf

of Edinburgh and Heriot-Watt University – represents the largest ever single UK public sector investment in the academic sector (Data Driven Innovation Programme, 2022).

Figure 10: Bayes Centre Exterior | Source: The University of Edinburgh

Figure 11: Advanced Computing Facility at the University of Edinburgh's Easter Bush Campus | Source: FutureScot.com

Figure 12: Former Prime Minister Boris Johnson visiting the Edinburgh Parallel Computing Centre | Source: University of Edinburgh

2.1.2. Summary and justification of case selection

Edinburgh, in its long history, has been subject to major urban reform. Most famously, it transformed and reinvented itself from "auld reekie" to enlightened New Town. Starting in the second half of the 18th century, it became a major intellectual centre, earning it the nickname "Athens of the North" because of its neo-classical architecture and reputation for learning. The CRD constitutes one of the most important urban redevelopment projects in Edinburgh since the development of the New Town in the 18th century. Contrary to previous developments efforts such as the notorious tram project, the BioQuarter[8] or St James Quarter[9], the CRD constitutes a more holistic urban-regional development project (McCrone, 2022). Seeking to become the DCoE, it offers the unique opportunity to witness historical urban transformation first-hand. Adopting a purposeful sampling strategy, the CRD constitutes intensity sampling since the research phenomenon – projective agency in an urban planning setting – manifests itself in a rich and intense way (Patton, 2002). The CRD and

8 More information can be found here: https://edinburghbioquarter.com

9 More information can be found here: https://stjamesquarter.com

particularly its DDI Programme constitute two highly interesting units of observation. DDI is promised to be *the* key driver of economic growth. By definition, innovation undermines repetition, and therefore "dictates a crucial role for contingent imaginaries and social representations of uncertain futures" (Boyer, 2018, p. 42) into the implementation process of the CRD.

Figure 13: The Edinburgh and South-East Scotland City Region Deal | Source: Own illustration

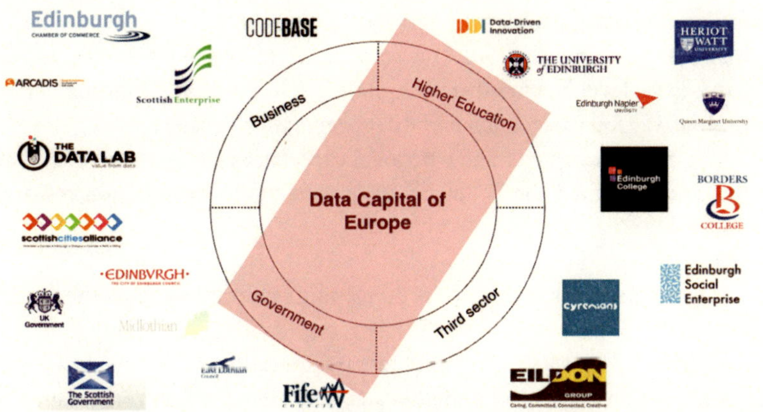

The CRD embodies the fusion of urbanisation[10] and datafication. It reflects the dominant regime of the UK's subnational economic development policy. It allows us to track and observe the implementation of DDI as a means of tackling big issues, such as slow growth and social inequality. It aligns a cluster of extremely powerful actors – political leaders, higher education officials, tech entrepreneurs – around one specific sociotechnical imaginary – establishing the city region as the DCoE (Figure 13). Put differently, it provides the opportunity to witness the production of a system of city-making in its nascence, before it becomes solidified into (digital) infrastructure.[11] Given the research

10 Edinburgh's population is set to grow to 600.000 in the next 20 years https://worldpopulationreview.com/world-cities/edinburgh-population

11 It is important to emphasise that "solidification" and, subsequently, "infrastructure" does not imply stasis but is rather marked by an inherent degree of change and dynamism.

problem at hand and the open and real-life nature of the CRD, it is appropriate to design it as a single case study (Yin, 2008).

2.2. Data generation

Having introduced the empirical case study, I now want to elaborate more thoroughly on the methodological strategy: the logic by which I approached the research questions at hand. The development of the research problem expresses a mechanical intellectual puzzle. I was interested in how something works or is constituted (Mason, 2002). That "something" was projective agency in an urban planning setting. After finding out about the CRD and selecting it as a single case study, I officially started my empirical fieldwork in November 2019. In what follows, I will first briefly reflect on the guiding research questions which paves the way for a discussion of the data generation process.

Originally interested in how fictional expectations develop (in) the urban landscape, I entered the fieldwork stage with the following main research question:

1. How are urban planning agents developing expectations about the future?

Seeking to unpack this overarching research question, I developed two sub-questions, which built on one another and structured the data generation process:

2. What is the form and content of late modern, data-driven urban-regional futures?
3. How are urban-regional fictional futures practiced, individually and collectively, by whom and when?

The two sub-questions indicate two empirically distinct, yet dialectically related research paths. The first question addresses the *content and form* of fictional urban-regional futures. The second question is aimed at the *process* of imagining urban-regional futures. Importantly, the two research questions illustrate an analytical hierarchy. As a first step, I intended to build a detailed picture of what kind of fictional futures were being imagined by whom. This first question would then contribute to the second sub-question of this study; namely, the process of imagination. Here, I sought to examine how exactly urban planning agents construct and practice fictional futures. In the course of the fieldwork, a third sub-question was added related directly to empirical findings, which suggested the significant influence of scientific expertise in crafting and implementing economic futures for the Edinburgh city region. In addition to the form and content of urban-regional futures and the process of imagining them, I asked:

4. What role does scientific expertise play in shaping contemporary imaginaries of the datafied city (region)?

The four overarching research questions built the foundation for the data generation stage[12], which lasted from November 2019 until December 2020. Importantly, the four guiding questions constituted starting points for the empirical fieldwork and were constantly revised and specified (see Figure 14 for a summary). In Section 2.3., I will revisit the initial research questions and link them to the following empirical chapters.

Engaging with the underlying intellectual puzzle and the research questions that guide it, this research project required answers that were rich and in-depth. This called for methods that elicit context, multi-dimensionality and complexity rather than a broad understanding of surface patterns (Mason, 2002). The aims of the research questions were to study behaviour, practices and processes within selected socio-technical assemblages. This "interpretivist" approach differs from positivist approaches which seek to investigate the influence and dependence of one variable on another, based on rigid and standardised methods of data generation. While questionnaires can deliver wide and representative data, they often lack depth and broadness. This study was concerned with "how the social world is interpreted, understood, experienced, produced or constituted" (Mason, 2002, p. 3), and was therefore *qualitative* in nature.

12 Given that I am not a neutral collector of information about the social world, I will use the term data generation instead of data collection (Mason, 2002).

Figure 14: Units of observation and research questions | Source: Own illustration

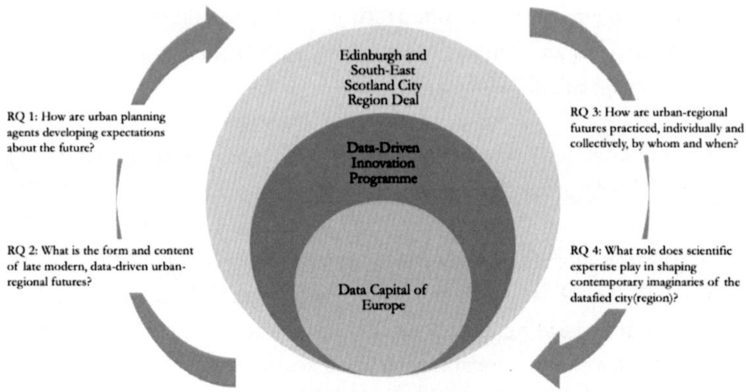

Importantly, the choice of methods must reflect the ontological assumptions guiding the study. Existing sociological accounts that empirically investigate perceptions of the future have adopted a variety of methods. Beckert and Suckert (2021) emphasise a shift in research methods from predominantly quantitative approaches in the 1960s (surveys and statistical analysis) to increasingly qualitative approaches in recent contributions. Mische (2014) points out that "future projections can be studied via their externalisations in attitudes, narratives, performance, and material forms" (p. 437). Finally, scholars in STS, such as Jasanoff *et al.* (2006), suggest comparison, interviews and document analysis as pre-eminent methods for the analysis of STIs. This book attempts to contribute to the increasing relevance of constructionist, interpretative, and culturalist approaches in sociology by triangulating in-depth interviews, documents and ethnographic elements.

In this study, urban-regional innovation is conceptualised via the vocabularies of STS where the city is understood as a "multiplicity of changing, co-existing and mutually interfering urban assemblages" (Blok and Farias, 2016, p. 2). Studying urban-regional assemblages – entanglements consisting of human and non-human actors – required a methodological approach that did not view methods as siloed and separated elements in the process of knowledge generation. Instead, it was crucial to adopt a reflexive approach that acknowledges the contingent, mutually dependent and *political* nature of methods. Rather than hiding behind methods as innocent techniques of knowledge production, it was important to take into account their performative nature. Law (2004) correctly asserts that methods "not only describe but also help to produce the reality that they understand" (p. 5). Adopting seemingly straightforward methods such as qualitative interviewing or document analysis, Law reminds us, of the tendency to simplify social complexity. Therefore, it was important to critically examine the chosen methods for the directions that they pushed me in, and also embrace, rather than reduce, ambivalence and multiplicity. In what follows, I will elaborate in more detail on the three research methods.

2.2.1. Documents

Documents constituted a highly relevant source of data, particularly in the initial stage of this research project. Freeman and Maybin (2011) assert that "the document is a translation that also translates" (p. 160) since it "mediates between decision makers, practitioners and their publics" (p. 165). This insight is crucial because it demonstrates that documents are imbued with power relations. Not only does it matter who *writes* documents but it is equally important who *reads* them and *talks* about them. In many ways, documents act as "boundary objects" (Star and Griesemer, 1989). Key documents such as the official CRD document, signed in August 2018, "have different meanings in different social worlds but their structure is common enough to more than one world to make them recognizable, a means of translation" (Star and Griesemer, 1989, p. 393). The official CRD document ties together actors, discourses and spaces. It provides a constant reference point by guiding action in the present and building the foundation for future activities.

Keeping in mind the multi-scalar nature and strong interconnections of politico-economic imaginaries, I selected documents from regional, national and global actors. Designing the research, public documents were already drawn upon in order to gain a first understanding of the research puzzle and identify new areas for analysis. Conducting document analysis, a total of 48 documents were reviewed (see Appendix, Figure 38). These included media reports and analyses, corporate and governmental announcements as well as academic contributions such as speeches, minutes, press conferences, papers and reports. Examples include the UK government's *Industrial Strategy 2017*, the City of Edinburgh Council's *Economic Strategy 2018* or the OECD's report *Supporting Investment in Knowledge Capital, Growth and Innovation*. Most of these documents were publicly available. Therefore, access did not pose a problem.

Examining these documents, it was important to keep a critical mind; that is, acknowledge that documents, whether visual or textual, are constructed in particular contexts by particular people with particular interests (Mason, 2002). For instance, the Science and Innovation Audit of the Edinburgh city region (UK Government, 2017) – a precursor to the official CRD document – differs starkly in context of emergence, tone and expected audience from the Audit Scotland report (2020). The former – written by a local consortium of higher education officials and council managers – praises DDI as the key driver of future economic growth in the Edinburgh city region. The latter document, commissioned by the Scottish Government's public spending watchdog for local government, criticises plans to develop the city region into the DCoE for its lack of citizen participation. The two examples underline the importance of keeping in mind "who is allowed or tasked to write what and by whom" (Freeman and Maybin, 2011, p. 163).

2.2.2. Interviews

Building on document analysis as an initial source of data, in-depth interviews were the primary research method of this study (see Appendix, Figure 39). Given that this research project raised questions that documents could not address in enough detail, and the fact that the period under study is in the now, interviews with key decision-makers involved in crafting the DCoE vision were an extremely important source of data. Kvale (2007) points out that the qualitative interview "provides a unique access to the lived world of subjects, who in their own words describe their activities, experiences and opinions" (p. 9). Keeping the qualitative nature and the ontological and epistemological assumptions underlying this research in mind, a stratified purposeful sampling strategy was adopted. This type of sampling allowed for the selection of varied and information-rich cases, thereby permitting to capture commonalities and variations among the chosen samples (Patton, 2002).

This research project studied the practice of imagining futures as an essential part of urban assemblages through specific actors. Seeking to gain an in-depth understanding of the construction and dispersion of the DCoE imaginary, a total of 50 interviews were conducted. The selection of participants was closely related to the four main research questions and the sub-questions that underpin them. Importantly, different interviewees required the development of different interview schedules. While the interview is commonly seen as a "conversation with a purpose", it is important that the decision of who to interview is related to the intellectual puzzle, and that the interview questions are grounded in the research questions (Hooley, Marriott and Wellens, 2012). Although there were common elements across all interview schedules, such as a focus on open-ended questions and a content-related move from the abstract to the specific, questions posed to business representatives differed from questions posed to academics.

All of the interviews were carried out in Edinburgh in no particular order. However, given the fluid, reflexive and data-driven character of qualitative research, it was important to ensure that a sufficient degree of temporal overlap occurred between the different interview sets. As a result of the Covid-19 pandemic, most of the interviews (38 of 50) were conducted via online video calls using Microsoft Teams or Zoom. Each interview was recorded by use of an Olympus VN-541PC recorder. The interviews were carried out synchronously. They "mirrored a traditional interview in that they took place in real time but in an online environment" (James and Busher, 2012, p. 3).

There are advantages and limitations to this form of data generation. Besides obvious limitations, such as internet connection issues or poor video quality, the greatest challenge in this seemingly disembodied form of interviewing was the loss of subtle, non-verbal cues which, in a face-to-face scenario, can help to build rapport and gain the trust of the interviewee (O'Connor *et al.*, 2008). However, it is important to note that online interviews are not fully disembodied experiences. Rather, they are characterised by a technologically mediated form of embodiment. Hine (2015) points out that "being online" is not necessarily a "discrete form of experience but often [...] an extension of other embodied ways of being and acting in the world" (p. 41). It follows that conversing via video conferencing tools, such as Microsoft Teams or Zoom, does not portray an out-there experience but has become an intimate part of our everyday lives. The biggest advantages include the saving of cost and time and, in many cases, a more "informal" engagement, leading participants to be more open with others than in real-world communication (Joinson and Paine, 2012). This was predominantly due to the fact that interviewees would join the conversation from their homes.

Seeking to obtain rich data from the different participants, a typical interview unfolded in the following way. Moving from the general to the specific, I started by asking questions about the personal and professional trajectory of the respondent, the role of the organisation within the CRD as well as asking about the history of the organisation and the participant. These introductory questions then set the stage for inquiries into participants' understanding of data-driven, urban-regional futures. In other words, it was of interest to follow the different dimensions of projectivity, thereby generating data with regard to what data-driven futures consist of and what role the participant assumes in actualising them, both individually and organisationally. A second set of questions referred to how data-driven futures are governed, by whom and when. Particular attention was paid to how sociotechnical *imaginaries* relate to sociotechnical *realities*. An example of this was the juxtaposition of contrasting views of higher education and third sector representatives regarding the societal benefits of DDI.

Finally, access presents one of the central issues when conducting qualitative interviews. Access can be defined as "the process by which a researcher and the sites and/or individuals he or she studies relate to each other, through which the research in question is enabled" (Riese, 2018, p. 3). Following the initial pilot study, I made use of a snowball sampling technique (Patton, 2002), building on a preliminary list of potential interviewees as well as asking interviewees for assistance in identifying further potential interview participants. At the end of each interview, I would ask the participant whom they would recommend me to speak next to gain a better understanding of the design and implementation of the CRD. This strategy proved highly useful since I was able to get in touch with individuals who either had been very influential but were no longer part of the project or who were not part of the project at all but could provide valuable details on related aspects. As a consequence, I was able to connect to additional knowledge networks and gain a more nuanced understanding of this particular urban-regional development project.

2.2.3 Ethnographic vignette

Finally, I made limited use of ethnographic fieldwork (see Appendix, Figure 40). The epistemological position of this study suggested that knowledge or evidence of the chosen phenomenon could be generated by immersing myself in "natural" or "real-life" settings (Mason, 2002). Keeping the main research interest in mind, it was helpful to not only rely on participants' retrospective accounts but to put emphasis on situationally generated knowledge. To this end, I spent time with various stakeholders (smart city practitioners, academics, civil society) in order to develop a more nuanced understanding of the creation of smart urban-regional futures.

Due to the Covid-19 pandemic, the number of on-site observations was severely limited. I officially started my empirical fieldwork in November 2019. Until the Covid-19 outbreak, I undertook four on-site visits, including two smart city practitioner conferences in London and Oxford (Figure 15 and 16), one Joint Committee (key decision-making body of the CRD) meeting as well as one town hall meeting organised by the Cockburn Association (Figure 17). Most of the ethnographic fieldwork was conducted digitally. I participated in several online events, among others, Joint Committee meetings, policy conferences by the Westminster Higher Education Forum, "studio meetings" of the EFI and data conferences (e.g., the Scotsman Data Conference 2021).

Engaging with the production of "common urban worlds" and focusing on imagination as a vital practice or "mode of production" of urban worlds, Desmond's (2014) relational approach to ethnography was particularly helpful. Challenging place- or group-based ethnographies in sociology, Desmond advocates a relational approach, thereby giving ontological primacy to configurations of relations. He suggests "studying fields rather than places, boundaries rather than bounded groups, processes rather than processed people, and cultural conflict rather than group culture" (Desmond, 2014, p. 574). Consequently, I placed the *process* of imagining – the construction and diffusion of images of the future – at the heart of the research. While I did not conduct a full-fledged, systematic ethnography, I started to develop an understanding of how specific imaginaries such as the promise of transforming the city region into the DCoE align individuals, organisations, discourses and geographical spaces.

Approaching potential field sites, I considered two strategies or "roles" to be particularly useful. First, the role of the "historian", who intends to trace this major urban redevelopment project and second, the "expert", who is aware of similar urban transformation projects and wants to bring in sociological knowledge. Given that the University of Edinburgh represents a key stakeholder in the CRD, I was able to gain access relatively easy. Expecting to enter "sites of hyperprojectivity", meaning "communicative settings, somewhat removed from the flow of day-to-day activity, in which the explicit purpose of talk is to locate problems, visualize alternative pathways, and consider their consequences and desirability" (Mische, 2014, p. 447), I was struck by the mundaneness, lack of demonstrated reflection and banality of smart city thinking. One particularly memorable experience was a smart city conference in Oxford. Taking place in an oversized conference centre in the middle of an ordinary industrial park (Figure 16), the event started off with the famous Star Wars soundtrack, followed by a number of smart city vendors and entrepreneurial policy-makers presenting techno-centric visions of the smart city. I couldn't help but notice a slight discrepancy between participants' self-perception alongside envisioned urban futures and sociotechnical realities.

2.3. Revisiting the initial research questions

Having elaborated on the data generation process, I want to briefly revisit the initial research questions. The latter addressed 1) content and form of fictional futures, 2) the process of imagining data-driven futures, and 3) the role of scientific expertise in urban-regional development. Unsurprisingly, the empirical insights generated in 13 months of fieldwork brought themes and tensions to the surface that I had neither considered nor anticipated. Starting out with a strong interest in fictional expectations and the role of imagination in urban-regional decision-making, it became clear that this study sheds light on the interplay of scientific expertise and urban-regional planning. More specifically, it problematises the role of powerful HEIs such as the University of Edinburgh in shaping urban-regional economic futures.

As a result of the empirical fieldwork conducted, at least three tensions emerge: 1) Edinburgh's historical identity vs. its future-oriented ambitions around DDI, 2) the role of scientific expertise in shaping smart urbanism, and 3) technocratic vs. citizen-centric smart urban development. Consequently, the four overarching and rather descriptive research questions could be specified in the following way:

1. Does the investment in data and scientific expertise, intrinsic to the idea of the smart city, provoke problems for democratic accountability and legitimacy in urban governance?
2. Why are particular kinds of urban-regional knowledge such as data science given importance while other strands of knowledge are side-lined?
3. Which competing imaginaries of the modern university are at play when reconceptualising the Edinburgh city region as the "Data Capital of Europe"?
4. What role does technoscience play historically in re-centring the social? That is, aligning social orders around new or different understandings of urban-regional sociality and the social good?

In many ways, this explication of the research focus also reflects an intellectual shift away from an interest in the role of imagination and creativity in economic decision-making to an engagement with the interplay of science and urban-regional democracy. Entering the fieldwork stage, my thinking was heavily shaped by Hartmut Rosa's (2013) theory of social acceleration and Jens Beckert's (2016) notion of fictional expectations. Having conducted 50 in-depth interviews, analysed 42 documents and participated in a range of online and offline events, I was able to tailor the initial research questions to the empirical case study. As a result, I started to conceptualise smart urban development through the lens of STS literature at the intersection of science and urban studies (Beck, 1992; Jasanoff, 2004; Kitchin, 2014; Jasanoff and Kim, 2015; Pfotenhauer and Jasanoff, 2017; Zukin, 2020). The following empirical chapters (Chapters 4–7) will be testament to the empirical and conceptual insights gained during the research process.

Figure 15: Smart Cities 2020 conference London | Source: Own image

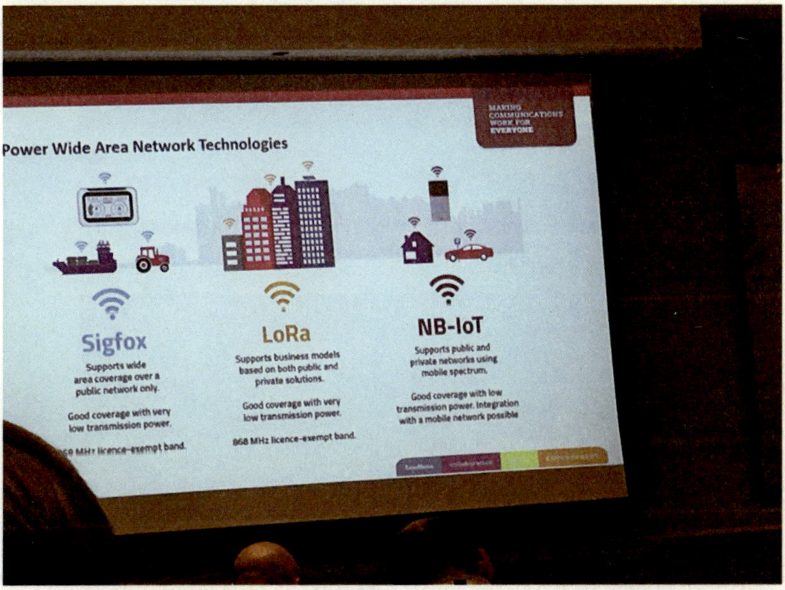

Figure 16: Expected "site of hyperprojectivity" | Source: Own image

Figure 17: Town hall meeting "City for Sale" by the Cockburn Association | Source: Own image

2.4 Data analysis

Given the qualitative and reflexive nature of this study, the data generation and data analysis were intimately linked. The analysis involved the transformation of often voluminous and varied data into a "a clear, understandable, insightful, trustworthy and even original analysis" (Gibbs, 2007, p. 1). It consisted of two main elements: data handling and data interpretation (Mason, 2002). Bearing in mind the three research methods, this study created large amounts of textual data manifested in the form of interview transcripts, field notes and collected documents.

In order to facilitate the organising of the generated data and support the identification of patterns within it, I made use of the computer-assisted qualitative data analysis software NVivo 12. The software was used to code interview transcripts, documentary and visual material. Concerning the processing of qualitative interviews, for instance, it is important to mention that transcribing in itself is an interpretive process (Kvale, 2007), which, while concerned with attention to detail, is part of the data analysis. In line with this, interview transcripts and field notes were treated as only possible transformed representations (ibid.). The transcribing was done by use of the AI-power audio transcription software Trint[13]. The switch from manual to AI-based transcription proved extremely valuable since I could save time transcribing while placing stronger emphasis on editing and revision.

Central to the processing of the generated data was the idea of thematic coding or categorising. This involved "identifying and recording one or more passages of text or other data items [...] that, in some sense, exemplify the same theoretical or descriptive idea" (Gibbs, 2007, p. 38). Given the exploratory nature of this research, I adopted a data-driven or open coding approach. This implied the creation of new codes, thereby not imposing pre-existing theoretical frameworks on the data. It goes without saying that a complete elimination of presuppositions is unrealistic. Nevertheless, it seemed useful to keep an open mind, and only in the later stages of the data analysis move back and forth between data-driven and concept-driven coding approaches.

---

13 For more information, see: https://trint.com

Figure 18: Summary of nodes | Source: Nvivo12

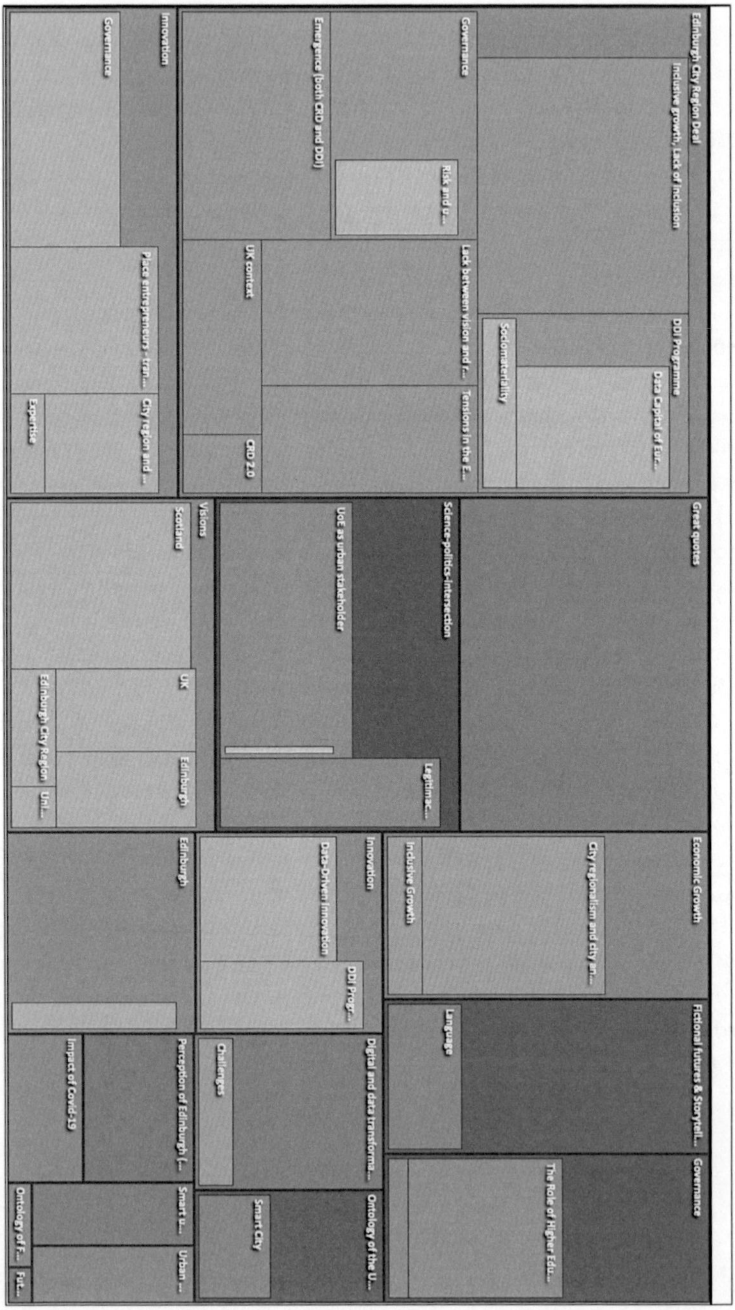

My coding strategy consisted of three parts: 1) fieldnotes, 2) descriptive coding, and 3) memo writing. Following the interview, I took fieldnotes to record immediate impressions and thoughts. This proved highly useful since it allowed me to reflect on emerging themes as well as the interview flow. In a second step and following the transcription process, I conducted descriptive coding in NVivo 12. I gathered related textual data and clustered them around "nodes" in order to identify and solidify emerging analytical themes (see Figure 18 for a summary). This strategy allowed me to develop a detailed and nuanced understanding of key themes such as DDI or the role of legitimacy along the science-politics intersection.

Finally, I created memos for each of the 50 in-depth interviews.[14] Charmaz and Belgrave (2012) underline that "memo writing links coding to writing the first draft of the analysis" (p. 15). Building on the identified nodes, I "fractured" the data by 1) defining the properties of each category, 2) specifying the conditions when the category developed, and 3) delineating its consequences and relationships with other categories (ibid.). For instance, based on the node "science-policy intersection", I identified "boundary work" (Gieryn, 1983) as an important analytical category. This category, in turn, is supported by evidence from data (i.e., illustrative interview excerpts).

2.5 Ethical considerations and risk assessment

Finally, it was vital to take into account ethical considerations before, during and after the data generation process. Acknowledging and dealing with power relations, ensuring confidentiality and developing a critical awareness of my own positionality seemed particularly salient issues in the context of this research project.

Concerning power relations, this study was generally dealing with research participants, who were more powerful and held a higher socio-economic status than the researcher. Three immediate implications followed. First, I needed to be aware of the fact that research participants may want to control and dominate the interview (Richards, 1996). For instance, when confronting a senior university official with the increasingly entangled private-public relationships that characterise data-driven social reforms, they reacted aggressively and, in turn, attempted to lecture me about the apolitical nature of HEIs. This inci-

14 The template can be found in the Appendix (Figure 42).

dence highlighted the importance of not only being well prepared but also triangulating interview data with other sources of data.

Second, it was crucial to keep a critical distance to my object of analysis. However, this proved more difficult than initially assumed. Being a member of the University of Edinburgh myself and engaging with urban-regional decision-makers on a regular basis, I acquired a form of "interactional expertise" (Collins and Evans, 2007), thereby occasionally lapsing into a "cheerleading tone". Examples of the latter include the uncritical use of terms such as DCoE, "place entrepreneurs" or "world-class research". Seeking to preserve my critical distance, it was important to keep in mind the performative power of language and triangulate interview data with other sources of data (e.g., the Audit Scotland (2020) report on Scottish city region and growth deals).

Third, interview participants could refuse to sign a written consent form and, additionally, have legal teams in the background who would protect them in case of any legal or ethical infringement. Therefore, I expected to obtain verbal consent in the case of most of the interviews. Besides sending interviewees an information and consent sheet before the interview, I made sure that interviewees agreed to being recorded. In addition to dealing with power relations, it was crucial to ensure the anonymity of the interviewees. This involved the use of pseudonyms wherever possible and informing participants about who would have access to the data. In the event that anonymisation could not be preserved, for instance in the case of senior officials who are easily identifiable, I discussed issues of confidentiality in each case and agreed with them on the publication of quotes. It goes without saying that any requests for non-publication of any data were adhered to.

Lastly, it was highly important to develop a critical awareness of my own positionality. Given the significance of the CRD and its politically sensitive nature, I needed to be aware of potential conflicts of interest (e.g., the witnessing of corruption) occurring during the fieldwork process. For instance, one interviewee asked me to turn off the recorder at one point during the conversation, only to tell me about illegal activities occurring at one of the partner organisations. In addition to that, it was important to be aware that major urban redevelopment projects such as the CRD may be impacted by overall changes in the political system of the UK and in Scotland. For example, it remains highly interesting to observe how continuing independence efforts by the Scottish Government shape the role of the Edinburgh city region.

3. Conclusion

In this chapter, I have introduced the research design underlying this study. This was done by tracing the research journey from project development to data analysis. There are three main takeaways, which pave the way for the subsequent empirical chapters. First, the Edinburgh CRD constitutes a highly significant empirical case. As the first higher education-driven city region and growth deal in the UK, it offers fascinating insights into urban-regional planning at the intersection of science and democracy. It allows us to study how a range of powerful agents – higher education managers, government officials, tech entrepreneurs – attempt to bring into being desirable politico-economic futures for the Edinburgh city region.

A second lesson learned from undertaking the research refers to the inherent contingencies involved in conducting empirical fieldwork. Arriving in Edinburgh in September 2018, I was unaware that the Edinburgh CRD had been signed just a month before. By coincidence (and a degree of curiosity), I stumbled across the EFI. Finally and following a range of pilot interviews with representatives from the EFI, I decided to focus on the CRD as my primary case study. Importantly, this research projectstudies a specific moment in time. The CRD spans 15 years. My study followed the CRD from its emergence in August 2018 until the end of 2020. It goes without saying that subsequent developments shaped the analysis. However, the majority of the empirical findings were obtained between November 2019 and December 2020.

Third, the data generation process revealed the importance of keeping some degree of critical distance to the research objects and the performative and inherently exclusionary nature of methods. Regarding the former, being a member of the University of Edinburgh – the key driver of the CRD – I had to carefully consider my role as a researcher and my position as a PhD student. It was paramount to keep in mind that methods do not constitute innocent or purely technical procedures in the process of knowledge generation. Instead, they help craft and enact particular realities (Law, 2004). The selection of methods (interviews, documents, ethnography) gives voice to specific presences while silencing others.

In the next chapter, I will trace the construction of Edinburgh's innovation assemblage. Bringing 13 months of empirical fieldwork into conversation with existing academic perspectives, I am going to shed light on who shapes futures of the Edinburgh city region and to what effect.

Chapter 4: Constructing Edinburgh's Innovation Apparatus

1. Introduction

Who shapes urban-regional futures and to what effect? How did the DCoE imaginary emerge? How was it embedded into existing ideas and imaginaries about the city region of Edinburgh? In this chapter – the first empirical one of this book – I trace how imagination is turned into social practice. I describe how the Edinburgh City Region Deal (CRD) emerged from a conversation in a pub in Southern France to a significant £1.3bn investment aiming to transform the Edinburgh city region into the DCoE.

In doing so, I conceptualise urban-regional innovation through the lens of STS literature at the intersection of science and urban studies. Rooted in the co-productionist tradition of science and technology studies (STS) (Jasanoff, 2004), which acknowledges that technoscientific and sociopolitical orders are mutually constitutive and dependent, I combine Jasanoff's and Kim's (2009, 2015) analytical concept of "sociotechnical imaginaries" (STIs) with the analytical vocabularies of urban assemblage thinking (Farías, 2011; Färber, 2014, 2019; Farías and Blok, 2017; Davis and Groves, 2019; Schneider and Lösch, 2019).

I argue that Edinburgh's innovation apparatus can be understood as a "promissory assemblage" (Färber, 2019). The concept of "promissory assemblage" allows scholars to emphasise the promise-oriented elements of cities as key coordination devices. The idea of transforming the city region into the DCoE aligns actors, organisations, discourses and geographical spaces and functions, as a panacea that promises eternal economic prosperity. Developing this line of argument, I conclude that Edinburgh's innovation apparatus comes into being through 1) three policy changes, 2) the possibility to mediate between local, national and global identities, and 3) the mobilisation of human and non-human agents of extension.

This chapter pursues two objectives. First, to contribute to ongoing debates at the intersection of STS and urban studies by shedding light on the production of an urban-regional sociotechnical imaginary. Second, to describe Edinburgh's innovation apparatus by teasing out empirical tensions, such as the unequal distribution of projective agency, which will lay the foundation for subsequent discussions in the remainder of the book. To this end, the chapter is divided into three sections. I begin by introducing the theoretical apparatus used for conceptualising urban-regional innovation in Section 2. After that, I decipher Edinburgh's innovation apparatus by means of three interlinkages: scales, identities and agencies in Section 3, and concluding remarks follow in Section 4.

2. Conceptualising urban-regional innovation

This study aims to shed light on projective agency in an urban planning context. Taking projective agency as the primary unit of analysis not only allows to circumvent the speculative nature of potential future scenarios, since it emphasises the work involved in creating futures, but most importantly, meets the requirements of analysing the complex interplay of temporality and social action. At this point, it is important to clarify that economic actors or, more specifically, urban planners, are never confronted with the binary choice between either ergodic, and therefore fully predictable, or completely non-ergodic, futures. In fact, one of the aims of this book is to critically examine the adoption of uncertainty as a conceptual starting point for a sociological analysis of urban-regional innovation economies (see Chapter 5). For now, however, I will provide insight into the production and dispersion of a specific sociotechnical imaginary – transforming the Edinburgh city region into the DCoE.

Given that the primary unit of analysis of this research project – projective agency in an urban planning context – is characterised by an inherent degree of contingency, dynamism and uncertainty, the notion of "assemblage" or "assembling" (Latour, 2005) is particularly helpful. Rather than conceptualising and examining urban-regional transformations such as the CRD through the lens of neoliberal urbanisation, this chapter adds to ongoing discussions between assemblage thinking and urban studies by carving out interlinkages between individuals, organisations, discourses and geographical spaces.

While highly appealing as a contextual wallpaper, the critique of neoliberalism limits the analytical repertoire to documenting the role played by private sector actors and the promotion of competitiveness. It does so without delving into the complex mechanisms that make certain forms of governmental intervention possible and which shape them in their execution. As a result, I will substantiate Jasanoff's and Kim's (2009, 2015) high-level concept of STIs via the analytic vocabularies of assemblage urbanism (Farías, 2011; Färber, 2014, 2019; Farías and Blok, 2017; Davis and Groves, 2019; Schneider and Lösch, 2019). Before diving into empirical detail, however, it makes sense to briefly clarify some key ontological and epistemological assumptions underlying this study.

2.1. Imagining urban-regional futures

Seeking to bring together "the idealistic collective imaginations identified by social and political theorists and the hybrid but politically neutered networks or assemblages with which STS scholars often describe reality" (p. 19), Sheila Jasanoff and Sang-Hyun Kim define STIs as

> collectively held, institutionally stabilized, and publicly performed visions of desirable futures, animated by shared understandings of forms of social life and social order attainable through, and supportive of, advances in science and technology (Jasanoff and Kim, 2015, p. 4)

The notion of STIs provides a highly intriguing epistemological prism since it allows to bring together ANT's focus on analytical symmetry and distributed agency with the normativity and inherently political nature of the imagination. Bringing social and political theory into conversation with materially-centred accounts, such as actor-network theory (ANT), the concept of STIs puts forward an understanding of imagination that, on the one hand, acknowledges human agents and institutions as central actors while, on the other hand, noticing that its implementation requires the orchestration of intricate networks (Jasanoff and Kim, 2015). Jasanoff, Kim and Sperling (2006) point out that STIs "have the power to shape technological design, channel public expenditures, and justify the inclusion or exclusion of citizens with respect to the benefits of technological progress" (p. 2). This book is rooted in the co-productionist (Jasanoff, 2004) tradition of STS and explores the construction and diffusion of a specific *urban-regional* sociotechnical imaginary – establishing the Edinburgh city region as the DCoE.

A further aspect that is highly relevant when analysing how social action is shaped by forward projections is the role of materiality. Elaborating on the relationship between imagined futures and materiality in an urban planning context, there appear to be at least two dimensions: an outcome and a process dimension of materiality. The first dimension points to finalised or completed imagined futures, as exemplified by architecture. The second dimension hints at materiality within the imagination process itself. Here, one could think of objects such as drawing boards, computer-based visualisation programs or business plans. Given the main research interest of this study, the processual dimension of materiality is particularly relevant. This dimension points to ontological tensions between Beckert's (2009, 2013b, 2016) economic sociology of imagination and practice theory and, more specifically, its posthumanist strand, ANT.

Comparing the two approaches, while theorists such as Bruno Latour and Michel Callon include and acknowledge non-human actors as equally important co-producers of social action, it seems that scholars such as Jens Beckert ultimately place the human and human interactions as the determinant actor and action respectively. Nonetheless, ANT's preoccupation with hybridity, mixtures and complexity reveals its own shortcomings. Jasanoff (2015b) points out that ANT "risks a kind of moral nihilism, making all actions and agents seem equally empowered, or disempowered, and therefore equally responsible, or irresponsible, for the networks within which they function" (pp. 16–17). Rather than simply assuming the existence of actor-networks, it is highly important to investigate how human values and decisions have brought together elements of the network in the first place.

This study recognises that it is ultimately only humans who have the power to imagine. At the same time, the successful implementation of specific imaginaries requires the mobilisation and orchestration of sociotechnical structures. Consequently, Jasanoff's and Kim's (2009, 2015) concept of STIs is highly useful since it combines the materiality of networks with the normativity of politics. STIs can be understood as "mechanisms of co-production" (Jasanoff and Simmet, 2021), which elucidate meaning-making in at least two salient ways. They call for "(1) an integrated examination of societies across scales and domains of representation and social practice [...] and (2) a symmetrical inquiry that asks how knowledge and materiality are co-produced with social values and normativity" (ibid., p. 2).

Ultimately, the CRD is driven by actors at the intersection of science and politics. These human agents – policymakers, higher education officials and tech entrepreneurs – put Edinburgh's innovation apparatus in place. Their actions are imbued with norms and values about what economic futures ought to be attained and what futures seem less desirable. As a result, their imagination becomes inherently political. In the Edinburgh city region, this discourse centres around how to make use of big data and AI in order to drive economic growth and ensure innovation. Importantly, the futures envisioned by human agents become translated into urban-regional sociotechnical structures. It is here where the analytical vocabulary of urban assemblage thinking becomes useful since it underscores the interplay of human and non-human agents in shaping urban-regional futures.

2.2. What about the urban?

The imagination of data-driven futures is intimately linked to urban space. Cities constitute materialised imagined futures and sites for the imagination of novel futures (Amin and Thrift, 2016; Van Driessche, Ache and Lagendijk, 2017; Zeiderman, Dawson and Palmer, 2021). Future imaginations and expectations about technologies are increasingly interwoven with urban futures. The discourse and imaginary of the "smart city" (Kitchin, 2014b; Vanolo, 2014; Karvonen, Cugurullo and Caprotti, 2019; Kitchin, Cardullo and Di Feliciantonio, 2019) depicts the most recent example. Seeking to conceptualise urban-regional innovation, it is therefore crucial to establish a clear ontological understanding of the urban.

In this book, I engage with "the urban" by adopting an ANT or assemblage-thinking approach[1]. Put differently, "the city" is understood as a "multiplicity of changing, co-existing and mutually interfering urban assemblages" (Blok and Farias, 2016, p. 2). This initial definition entails three major implications: First, the urban is understood as an "emergent quality of the multiple assemblage process" (Farias and Bender, 2010, p. 15). That is, the city is not an "out-there reality" (ibid.). Second, urban ecology is assumed to be co-produced by human and non-human actors. Finally, cities are not perceived as bounded units or stable objects but rather as existing in multiple, overlapping ways (Amin and Graham, 1997; Latour and Hermant, 1998; Amin and Thrift, 2002; Law, 2002; Latour, 2005; Farias and Bender, 2010; Farías, 2011). In other words, there is not one Edinburgh; there are multiple "Edinburghs" within Edinburgh, which together comprise the city. For instance, there is an Edinburgh on a map, in a

1 I am aware of differences between ANT and assemblage thinking. Given striking similarities between the two approaches (Müller and Schnurr, 2015), I will use the terms interchangeably.

data plan, in a brochure, as lived and experienced by plural demographic constituencies, as a tourist image, as a datafied extrapolation or as an AirBnB location.

The notion of assemblage urbanism began receiving increased attention in 2011 when a polarising debate in the journal "CITY" erupted. Challenging existing ontological assumptions about the urban, scholars such as Farías and Bender (2010) and McFarlane (2011) argued against studying the city as an instance of something else – widespread processes such as capitalism, neoliberalism or urbanisation, for example. Critical urbanists such as Brenner *et al.* (2011), in turn, asserted that assemblage theorists reject structure in favour of "naïve objectivism" and silence asymmetries, inequalities and hierarchies. Farías (2011) responded by foregrounding that the adoption of an symmetrical ontology "involves unveiling the actual practices, processes, socio-material orderings, reproducing asymmetries in the distribution of resources, of power and of agency capacities" (p. 370), and therefore allows for a radical rethinking of political practices – a "democratization of democracy" (ibid., p. 371). Consequently, assemblage-based approaches to urban analysis challenge the primacy of individual actors such as urban planners and designers, or historical structures such as neoliberal capitalism. Instead, they recognise the active role of a wide range of social and material elements in planning practices (Davis and Groves, 2019).

Following the urban assemblage debate in 2011, a growing number of scholars apply ANT and the related notion of assemblage to urban studies. Particularly relevant in the context of this chapter are studies that explore the potential of ANT and assemblage thinking to gain a better understanding of planning processes (McGuirk, 2000; Rydin, 2013; Alvial Palavicino, 2015; Groves, 2017; Van Driessche, Ache and Lagendijk, 2017; Davis and Groves, 2019; Schneider and Lösch, 2019). Three key insights are worth highlighting. First, futures are produced in present planning not only in the form of concrete visions and imaginaries but also via "anticipatory practices" (Alvial Palavicino, 2015), such as cost-benefit analyses, which shape representations of futures (Van Driessche, Ache and Lagendijk, 2017). Second, urban planning processes can be conceptualised via the notion of "anticipatory assemblage" (Alvial Palavicino, 2015; Groves, 2017). The latter "accounts for the multiple ways in which anticipatory practices and expectations are arranged, and for performative effects" (Alvial Palavicino, 2015, p. 162). Finally, the notion of "anticipatory assemblages" highlights the political dimension of anticipatory urban action. In line with the overall argument of this book, scholars within the sociology of anticipation underscore that "the capacity to imagine and enact the urban differently" is "unequally distributed" (Davis and Groves, 2019, p. 15).

At this point, it is important to clarify that adopting a symmetrical ontology of the urban does not preclude the exploration of unequal agencies. In other words, recognising the wide range of social and material elements in urban planning does not necessarily imply equal distribution of projective agency. Avoiding ready-made explanations and developing a sensitivity for the interplay of human and non-human actors helps to shed light on how asymmetries, hierarchies and inequalities are (re)produced in the first place. This insight is crucial and will be further explored in Chapter 5 when I will focus on the "Gang of Four" – four senior officials of the University of Edinburgh responsible for crafting the DCoE vision.

2.3. Innovation *apparatus*

I want to further complicate these remarks on ontology and epistemology by engaging with the following question: how can one grasp the interplay of actors, organisations, discourses and geographical spaces involved in bringing into being visions of desirable urban-regional futures? In addition to outlining my ontological understanding of the urban and the epistemological prism I mobilise in the study of the latter, it is crucial to conceptualise the social construction of tech space more directly. In what follows, I suggest theorising the interplay of scales, identities and agencies – key analytical prisms guiding this chapter and book – through the notion of "innovation apparatus".

Scholars in STS, sociology, political economy and urban studies have conceptualised the relationship between technology, innovation and the urban via a range of epistemological lenses. Harvey Molotch (1976), for instance, introduced the notion of the "city as a growth machine" in order to shed light on the role of local business and political elites in shaping urban development. Examining post-Fordist American cities, Molotch argues that so-called "place entrepreneurs" are particularly important for urban reform since they have control over land which would increase in value as a result of growth. Put differently, coalitions of land-based elites utilise urban politics and policy in the pursuit of the accumulation of capital and wealth. Molotch (1976) points out that "an elite competes with other land-based elites in an effort to have growth-inducing resources invested within its own area as opposed to that of another" (p. 309). While advancing urban theory at the time, this conceptualisation of urban political economy reveals weaknesses since it exaggerates the role of property markets, neglects the role of the state in shaping urban-regional development and puts too much emphasis on human agency (MacLeod and Goodwin, 1999; Farahani, 2017; Anttiroiko, Laine and Lönnqvist, 2020).

Another prevailing paradigm in urban governance literature – also drawn from political economy – represents urban regime theory (Stone, 1989, 1993; Davies, 2002). According to Stone (1989), an urban regime constitutes "the informal arrangements by which public bodies and private interests function together in order to be able to make and carry out governing decisions" (p. 1). Emerging in the context of US urban politics, regime analysis "views power as fragmented and regimes as the collaborative arrangements through which local governments and private actors assemble the capacity to govern" (Mossberger and Stoker, 2001, p. 812). Challenging the proposition that governing capacity is easily captured through the electoral process, regime theory stresses the importance of public-private collaboration and advances an understanding of power that emphasizes social production rather than control. That is, "power to" act rather than "power over" others. While contributing to a more refined understanding of the way that power works in urban settings and encouraging a shift away from a narrow focus on the formal institutions of elected government, the concept reveals two main limitations. First, because of its broad appeal and applicability in a variety of settings, the concept fell victim to what Mossberger and Stoker (2001) call "concept stretching". That is, the more it was applied in multiple settings, the less explanatory power it held. Second, there is widespread agreement among scholars that urban regime theory represents one era of urban governance, but that contemporary eras demand new approaches (Judd and Laslo, 2013). Clarence Stone (2015), one of the founding fathers of the concept, himself notes that "the one-time high-cohesion regimes held together around a redevelopment agenda [...] have given way to more diffuse governing arrangements and a less sharply defined agenda (p. 102).

A further theorisation of urban-regional innovation economies that has received considerable scholarly attention represents what Etzkowitz and Leydesdorff (1995) call the "triple helix model" of innovation. Studying research centres like Silicon Valley the authors emphasise new institutional configurations between academia, government and industry, in which universities are perceived as "natural incubator[s]" (Etzkowitz, 2003, p. 325) driving knowledge-based economies. While providing helpful insights on the synergistic interplay of academia, industry and government, the "triple helix model" of innovation remains too simplistic and paints processes of innovation as being devoid of political intent. What is missing is a recognition of higher education institutions as allotropic[2] institutions that do not simply engage in the seemingly neutral production of knowledge but simultaneously make claims on how urban societies *ought* to develop (Pfotenhauer and Jasanoff, 2017a; Pfotenhauer and Juhl, 2017).

Sharon Zukin's (2020c, 2020d, 2021) recent work on the "innovation complex" complicates existing conceptualisations of the social construction of tech space by illuminating the multi-scalar nature of urban innovation. Tracing the emergence of New York City's tech economy since the 2008 global financial crisis, Zukin disentangles the complex interrelationships between institutional actors (tech firms, universities, venture capitalists and local government), geographical spaces (e.g., university campuses) and discourses, which bring into being the "innovation complex". Constructing the latter, she points out, "requires material structures of buildings and land, social structures to train a workforce, and financial mechanisms to integrate public- and private-sector capital investment and direct it toward tech production" (Zukin, 2020c, p. 947). Although sometimes lacking theoretical grounding and engagement with existing academic debates in sociology and STS, Zukin's "innovation complex" provides valuable insights for this book's exploration of urban-regional innovation economies.

Building on these existing theorisation efforts and the remarks made in Section 2, I suggest characterising Edinburgh's emerging tech economy with the term "innovation *apparatus*". This assertion necessarily requires justification. Rooted in a Foucauldian understanding of power[3], I contend that the notion of *apparatus*[4] incorporates the complexity of innovation while avoiding on-

2 A term used in environmental sciences denoting "the existence of two or more forms (allotropes) of the same element" (Park and Allaby, 2013). Allotropic institutions such as universities or museums are organisations that exist in many different states at once. That is, global higher education institutions such as Edinburgh University cater for plural constituencies and values – ranging from education, markets and consumption to civil society.

3 Foucault (1996) suggests a relational conception of power and defines the latter as "a whole series of particular mechanisms, definable and defined, that seem capable of inducing behaviors or discourses" (p. 394). Power is not held or exercised centrally by individuals but "operates through practices that act upon subject's present or future actions" (Miceli and Posada, 2022, p. 3).

4 The term "apparatus" constitutes the English translation of what Foucault, in his "The Confession of the Flesh" (1977) interview, initially called "dispositif" (Foucault, 1980).

tological indeterminacy and flux. In an interview, Foucault (1980) tentatively defined the notion of apparatus as

> a thoroughly heterogeneous ensemble consisting of discourses, institutions, architectural forms, regulatory decisions, laws, administrative measures, scientific statements, philosophical, moral and philanthropic propositions – in short, the said as much as the unsaid. Such are the elements of the apparatus. The apparatus itself is the system of relations that can be established between these elements. (p. 194)

Elaborating on the knowledge-power dyad, Foucault stresses that the apparatus has a "dominant strategic function", which implies a "manipulation of relations of forces, either developing them in a particular direction, blocking them, stabilising them, utilising them" (Foucault, 1980, p. 196). This initial conceptualisation of "apparatus" is helpful since it accounts for the co-productionist nature of knowledge and power while recognising the interplay of individual elements – linguistic and non-linguistic – in the pursuit of particular ideological projects. Applying this theoretical framework to the underlying empirical case study, it is the idea of innovation that drives actors, organisations, discourses and geographical spaces in a particular direction. Thus, the notion of the "innovation apparatus" can be defined as the network of discourses and imaginaries, practices, actors (human and non-human) and geographical spaces in urban-regional innovation, and the knowledge-power relationships inscribed and re(produced) between these elements.

Considering my ontological remarks on the urban in Section 2, it is crucial to briefly address my handling of the concepts of *apparatus* and *assemblage*. Rather than viewing the two notions as mutually exclusive conceptualisations of agencies, I agree with Legg (2011) who argues that "apparatus and assemblage [...] emerge as one and part of each other but in a continual dialectic" (p. 131). While coming from different conceptual genealogies (Deleuze and Foucault), both concepts emphasise relational ontologies, heterogeneity and constantly changing space-time relationships (Agamben, 2009; Legg, 2011; Müller and Schurr, 2016). Importantly, though, Legg (2011) contends that "apparatuses be considered a type of assemblage, but one more prone to (in the sense of anticipating, provoking, achieving and consolidating) re-territorialisation, striation, scaling and governing" (p. 131).

Therefore, against the background of these theoretical considerations and reflection on the interplay of theory and empiricism in the context of the Edinburgh city region, I start from a dialectical understanding of "assemblage" and "apparatus". In other words, the city is understood as consisting of a multiplicity of urban assemblages including amongst others, what I describe as "innovation apparatus". Supporting my conceptualisation of the relationship between technoscientific innovation and urban politics, it is the relatively weak autonomy of individual elements as well as the tendency for alignment and governing that is inscribed in the notion of apparatus. As a result, the concept of STIs – ultimately rooted in Foucauldian thought – helps to operationalise the idea of the innovation apparatus as a "system of relations" – objects, subjects, discourses, spaces – that *orders* urban-regional society.

3. Grasping Edinburgh's innovation apparatus

In what follows, I want to lay out how the sociotechnical imaginary (Jasanoff and Kim, 2015) of transforming the Edinburgh city region into the DCoE was *constructed*, *embedded* into existing sociotechnical realities and *extended* across various scales. I will make use of the analytical vocabularies of assemblage thinking and its "empirical sister-in-arms" (Müller, 2015) ANT. Conceptualising the DCoE imaginary as a "promissory assemblage", I am going to demonstrate how science and technology became "enmeshed in performing and producing diverse visions of the collective good" (Jasanoff, 2015, p. 11).

Importantly, I am describing a specific moment in time. Keeping in mind the transitory and dynamic nature of the CRD[5], I seek to disentangle this promissory assemblage by paying particular attention to the complex "relations of history and potential" (McFarlane, 2011). Adopting Färber's (2019) concept of "promissory assemblage" seems particularly useful for this purpose since it explicitly focuses on the "temporality and virtuality of urban assemblages" (Färber, 2019, p. 265). In other words, it gives analytical access to

> The promises attached to modernity that are articulated in urban infrastructuring; the elusiveness of promises that we confront throughout modernisation and urbanisation; and to the connection between urban opportunity (and misfortune), subjectivation and endurance as a mode of activity that does not facilitate change (Färber, 2019, p. 270).

---

5 The CRD spans fifteen years (2018–2033). My empirical fieldwork focussed predominantly on the emergence of the DCoE vision (i.e., 2018–2021).

Following Färber's (2019) conceptualisation of urban assemblages, I argue that the idea of transforming the city region into the DCoE aligns *actors*, *organisations*, *discourses* and *geographical spaces* and functions as a panacea that promises eternal economic prosperity. In order to substantiate this argument, I will now shed light on each of the four elements responsible for producing and performing Edinburgh's innovation apparatus. It is the interplay of these four elements that enacts the DCoE imaginary. Rather than grasping Edinburgh's innovation apparatus through these siloed elements, however, I want to focus on interlinkages. To this end, I am going to elaborate on *scales*, *identities* and *agencies*.

3.1. Scales

Urban-regional innovation initiatives such as the CRD are characterised by their multi-scalar nature. The construction, embedding and extension of specific "imaginaries of innovation" (Pfotenhauer and Jasanoff, 2017a) unfolds across local, regional, national and global scales. Global innovation frameworks, such as the "Silicon Valley model" or the "MIT model", shape national innovation strategies, which, in turn, co-produce local and regional imaginaries. In what follows, I want to trace the emergence of the DCoE imaginary by highlighting two important facets: 1) the changing relationship between science and society, and 2) the imaginative work involved in creating the "Data Capital of *Europe*" imaginary.

3.1.1. Science and society, state and market

The emergence of city region and growth deals in the UK and their increased emphasis on innovation – particularly data-driven innovation – reflects a change in focus with regard to the role of 1) cities and 2) science and innovation in driving social and economic prosperity. In order to better understand the motives underlying the CRD, it makes sense to briefly elaborate on broader historical shifts shaping UK science and innovation policy.

Science and technology have traditionally shaped Western nations' self-imagination and figured prominently in supporting promising futures (Jasanoff, 2017). The UK is not an exception. Following the BSE or "mad cow" scandal during the late 1990s, a House of Lords (2000) report on *Science and Technology* concluded that science-society relations were in a "crisis of trust". The committee suggested moving away from the "deficit model", which imagined publics as insufficiently knowledgeable, and instead "engage the wider public in dialogue about what science could and should be doing" (House of

Lords, 2000, p. 13). Subsequently, the Labour Government, led by Tony Blair, attempted to reconceptualise the relationship between science and society from "a combative encounter between knowledgeable experts and ignorant lay masses [...] to a collective exercise of citizenship in a participatory democracy" (Gregory and Lock, 2008, p. 1257). Most prominently, in July 2002 the government conducted an unprecedent experiment in public dialogue on the future of genetically modified crops and food in the UK called "GM Nation?". While the results of this exercise were received with mixed feelings (Smallman, Lock and Miller, 2020), it certainly reiterated the government's desire to improve science communication and underscore the role of science as a key driver of societal progress more broadly.

Contrary to previous conservative governments – particularly the one led by Margaret Thatcher from 1979–1990, which pursued a "hands-off", market-based approach to science policy – Tony Blair's New Labour demonstrated a "greater willingness to increase public spending on scientific research, and a greater interest in promoting the commercialisation of academic discoveries" (Mansfield and Owen, 2020, p. 16). Speaking to the Royal Society in 2002, Tony Blair announced a "new age of discovery" arguing for stronger collaboration between science and society. Particularly interesting are his remarks on the science-politics intersection. While keeping up the imaginary of scientific autonomy, Blair, at the same time, advocated a rapprochement between science and politics by pointing to the interweaving of scientific facts and political judgements in decision-making. Asserting that Britain's path to the future was "lit by the brilliant light of science", the former Prime Minister, in 2006, not only reiterated politics' dependency on science but also the need to ensure that investment in academic science was matched by increased benefits for the economy and society as a whole.

The significance of science and innovation was subsequently reflected in the reorganisation of government departments, the creation of Technology and Innovation Centres (now Catapult Centres) and sustained investment in future technologies – despite the Great Recession of 2007/2008. This "new industrial activism" (Mansfield and Owen, 2020) was taken further by the coalition government (2010–2015) and led, amongst other things, to the establishment of UK Research and Innovation (UKRI) – an umbrella organisation consisting of all research councils and Innovate UK, which directs research and innovation funding for the whole of the UK. Following the EU referendum in 2016, Theresa May committed to crafting a "modern" industrial strategy. Similar to previous governments, this move was marked by the reorganisation of government departments. This time, the Department for Business, Innovation and Skills (BIS) was merged with the Department of Energy and Climate Change (DECC) to form the Department for Business, Energy and Industrial Strategy (BEIS). Launched in 2017, May's *Industrial Strategy* put strong emphasis on innovation and comprised key policies such as the creation of the Industrial Strategy Fund[6], an increase in total R&D investment to 2.4 per cent of GDP by 2027 (Figure 19) and the establishment of the Transforming Cities Fund.

Figure 19: Gross expenditure on research and development | Source: Office for National Statistics and Commons Library Calculations

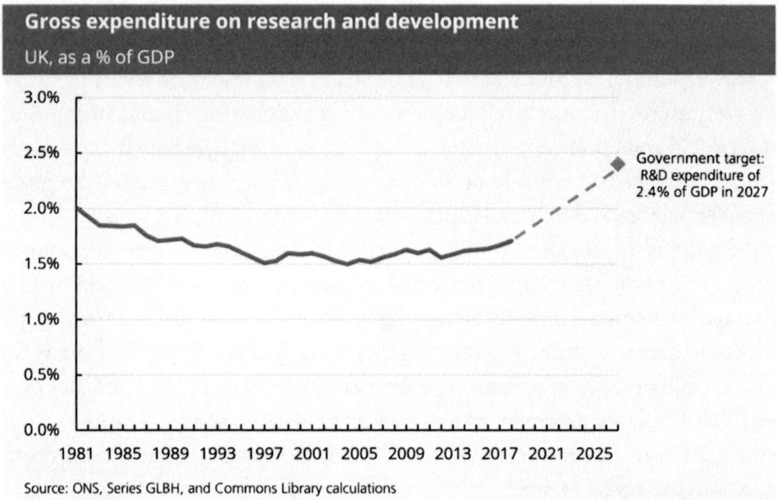

Fast forward to 2021, the Johnson Government appears to maintain Theresa May's approach to science and innovation. This is reflected in an increase of public funding for R&D to £22 billion per year by 2024/25 (UK Government, 2020), the establishment of a new visa regime to incentivise non-British scientists to come to the UK and the creation of a new funding agency – the

6 For more information, see: https://www.ukri.org/our-work/our-main-funds/industrial-strategy-challenge-fund/

Advanced Research and Invention Agency[7] (ARIA) – modelled after the US's Advanced Research Projects Agency (ARPA), to boost transformative research (Mansfield and Owen, 2020).

Crucially, since the late 2000s, the state assumes a more active role in innovation policy (Farmer and Gabriel, 2020). While UK innovation policy in the years following the Thatcher Government was characterised by the belief that "the market" is best placed to select promising innovations and disseminate its benefits, policy-makers increasingly take an interventionist approach, thereby actively shaping innovation. A good example of such action is the creation of the Future Fund[8], which aims at supporting innovative businesses affected by the Covid-19 pandemic. Here, the UK Government, in partnership with the British Business Bank, backs companies that are unable to access other government support programmes because they are either pre-revenue or pre-profit and typically rely on equity investment (British Business Bank, 2020). Initiatives such as the Future Fund or the recent announcement to invest £800 million in ARIA indicate the UK Government's strong commitment to science but more importantly, its ambition to assume a market-*shaping* rather than market-*fixing* role. Innovation – particularly data-driven innovation – is promised to secure global competitiveness and future economic growth. One organisational context that is particularly illustrative for gaining a better understanding of the interplay of science, innovation and society is the city.

3.1.2. City regions as drivers of economic growth

In order to understand the underlying dynamics shaping the CRD, it is important to recognise changing science-society relations and the increasing significance of innovation as driver of economic growth. Equally important is a critical engagement with the role of cities and city regions in UK and Scottish economic policy. One process is especially noteworthy: the creation of city region and growth deals.

As mentioned in Chapter 3, since the 1970s urban planning policy has been subject to frequent restructuring concerning governance arrangements and policies across different governments (Martin *et al.*, 2019). With the start of

7 For more information, see: https://www.gov.uk/government/publications/advanced-research-and-invention-agency-aria-statement-of-policy-intent/advanced-research-and-invention-agency-aria-policy-statement

8 For more information, see: https://www.british-business-bank.co.uk/ourpartners/future-fund/

the Cameron-Clegg Coalition Government in 2010, cities and city regions received increased attention as key drivers of future economic growth. This was reflected in the Localism Act 2011[9], which included, amongst other things, the *Core Cities Amendment*. The latter offered local councils in England the opportunity to submit plans on how to strengthen local economic growth and, if successful, negotiate deals with the UK Government for greater financial and planning autonomy (Ward, 2018). Subsequently, the idea was expanded by former Deputy Prime Minister Nick Clegg, who advocated a further shift of funding and decision-making powers away from central government to cities. What followed were several waves of CRDs and growth deals across the UK. As of April 2021, there are a total of 42 city region and growth deals across the UK.

In August 2014, the Glasgow and Clyde Valley CRD became the first city deal outside of England. Interestingly, up until this point, CRDs were not, or barely, mentioned in the Scottish Government's *Agenda for Cities 2011* and its *Economic Strategy 2015*. In 2016, the Scottish Government, in its refreshed *Agenda for Cities*, for the first time, outlines high level criteria for deals. These include increasing internationalisation, boosting investment and innovation and supporting inclusive growth. From 2016 onwards, CRDs became key vehicles for promoting urban-regional economic development. Currently, twelve city region and growth deals exist in Scotland (five signed and seven in the development stage). They are all multi-sector, multi-partner agreements with different thematic foci. For instance, Glasgow's CRD is predominantly focused on transportation, Aberdeen's revolves around energy and Edinburgh's is characterised by the strong involvement of higher education and its emphasis on data-driven innovation.

Compared to English CRDs, there are three distinctive features of Scottish economic policy that are worth mentioning. First, in 2011, the Scottish Cities Alliance – a collaboration of Scotland's seven cities and the Scottish Government – was established to promote Scotland's economic potential. The establishment of this intermediary body is unique since it indicates ambitions to find a Scottish-specific approach to urban development. Second, contrary to English CRDs, there is no transfer of power and decision-making to local regions in Scotland. Instead, the Scottish Government and the Convention of Scottish Local Authorities (COSLA) are working on a Local Governance Review, which aims to devolve more powers to local authorities (Scottish Government

9 For more information, see: https://www.legislation.gov.uk/ukpga/2011/20/contents/enacted

and COSLA, 2019). Finally, Scottish CRDs can be divided into two groups. A high-ranking policy manager within the Scottish Government points out that "there's the Glasgow deal and there's every other deal" (Interview 47, 2020). While the Glasgow deal resembles CRDs in the North of England and was developed without sufficient understanding of the long-term value and contribution of CRDs to the National Performance Framework, the deals signed from 2016 onwards are claimed to be better aligned with Scotland's economic strategy.

3.1.3. Smart urbanism and place-based innovation

The City of Glasgow not only constitutes the starting point for Scottish CRDs. It also plays a key role in the emergence of smart city innovation in the UK. In 2012, the UK Government's innovation agency, the Technology Strategy Board[10] (TSB), launched the Future Cities Demonstrator Competition (FCDC) to promote more efficient ways of integrating city services and systems. In the first phase, 29 municipal authorities were given £50,000 each to carry out feasibility studies to demonstrate how smart urban innovation could improve city governance. In a second phase, cities – based on their feasibility studies – submitted proposals for the £24m large demonstrator-funding (Arup, 2013). In January 2013, the City of Glasgow was announced as winner of the FCDC and awarded £24m to implement its proposal. In addition to Glasgow, the cities of Bristol, Peterborough and London received £3m each to bring forward elements of their proposals.

In many ways, the FCDC reflects some of the underlying tensions and conflicts characterising the role of smart city innovation in UK economic policy (Taylor Buck and While, 2017; Cowley, Joss and Dayot, 2018; Rose, 2019). Examining the FCDC, Taylor Buck and While (2017) highlight that effective public-private collaboration was prevented due to "short timescales, the need for projects to start quickly, the limited capacity of local government to forge appropriate links and, crucially, the friction between open source 'city gains' and closed source intellectual property-based profits" (p. 516). Glasgow's successful bid seems to be an exemplar for the technologically-biased and myopic type of bid, shaping the beginnings of smart urbanism in the UK. Revolving around the implementation of the Glasgow Operations Centre[11], the proposal put strong emphasis on a smart city management system, including web-

10 In 2014 rebranded as 'Innovate UK'.

11 For more information, see: https://futurecity.glasgow.gov.uk/ops-data/

based platforms, citizen engagement apps and digital infrastructures. A coherent, long-term approach to smart urbanism, including an understanding of "smart publicness" that goes beyond citizens as "entrepreneurs and service-users" (Cowley, Joss and Dayot, 2018) was largely missing.

Importantly, the FCDC points to the business-driven nature of smart urbanism in the UK. An Associate Director at Glasgow's Urban Big Data Centre underlines that smart urbanism "is more rooted in the business innovation and science and technology side of UK government" (Interview 3, 2019), having been "implemented through a series of demonstrator projects, a series of Catapult initiatives, a series of experiments, which [...] is different from more traditional planning-oriented approaches to urbanism" (ibid.). Rather than coming from the Ministry of Housing, Communities & Local Government, smart urbanism was advanced by BEIS and Innovate UK. One particularly important piece of the puzzle that helps connect the UK Government's identification of smart city innovation as key driver of future economic growth is the notion of place-based innovation.

In July 2015, then Universities and Science Minister Joseph Johnson announced the creation of Science and Innovation Audits (SIAs). Seeking to strengthen university-business relations and to promote local-regional innovation, SIAs are designed to "help local organisations map their research and innovation strengths and identify areas of potential global competitive advantage" (UK Government, 2015, p. 2). Consequently, local consortia consisting of higher education and research institutions, local partners and business were encouraged to submit audits to BEIS mapping local research and innovation strengths and infrastructure. One such consortium was formed in 2016, when senior officials from the University of Edinburgh and Edinburgh City Council identified data-driven innovation as key driver of economic growth for the Edinburgh city region (see Section 3.1.4.).

To summarise, the CRD represents the cumulation of three overlapping policy changes(Figure 20). First, since the early 2000s, science and innovation are framed as key drivers of economic growth with the state taking an increasingly interventionist approach. Second, the creation of CRDs, from 2011 onwards, has shifted the focus of economic policy from the state to cities and city regions. Finally, since the FCDC in 2012, cities are increasingly portrayed as sites and drivers of data-driven innovation. The CRD brings together science, innovation and the city. It embodies changing science-society relations, a strong belief in research and innovation, and an emphasis on cities and city regions as drivers of national economic growth.

3.1.4. "You just make it up" (Interview 25, 2020)

Having detailed three layers that illuminate the emergence of the CRD, I now want to shed light on the "imaginative work" (Jasanoff and Kim, 2015) involved in creating the DCoE vision. As mentioned in the previous section, one key policy responsible for driving local-regional innovation in the UK are SIAs. In 2017, a consortium of higher education officials and council managers conducted a SIA for the Edinburgh city region. Building on the *Shakespeare Review* (2013), which concluded that a second wave of economic and societal value will be generated by the "capacity to process and learn from data" (p. 5), the consortium envisioned the Edinburgh city region to become "a global destination of choice by 2025 for organisations powering services through the application of data science" (UK Government, 2017, p. 217). For the first time, Edinburgh was imagined as the "Data Capital of Europe" (UK Government, 2017, p. 8). How did the consortium arrive at this recognition?

The story begins three years earlier, in 2014, when a senior official at Edinburgh University commissioned an internal audit on the data capabilities of the university. An important finding from this audit was the realisation that the university does not own most of the data it holds and uses on a daily basis. As a result, it was suggested that the university should invest in its own data infrastructure by equipping the estate with sensors and turning it into an experimental data playground. Rather than limiting the sensor network to the university's estate, however, officials quickly began thinking about extending it to the entire Edinburgh city region. Utilising various sensor technologies (audio, infra-red, accelerometers) and multi-modal communication technologies, such as Wi-Fi, BLE, LoRaWAN and 5G,[12] the university hopes to engage in a wide variety of projects, including transport, health and tourism. Asked about the rationale behind a regional sensor network, one high-ranking employee in the University's IoT Research and Innovation Service pointed out:

> "This whole sensor network allows us to collect data, which we can then train new machine learning models to understand what's happening in the environment. And then we can deploy those models directly out to the edge

12 BLE – Bluetooth Low Energy – is a wireless technology that enables devices to be networked within an environment of about 10 metres. LoRaWAN – Long Range Wide Area Network – is a low-power wireless network protocol at the network layer level. 5G – fifth generation of mobile communications – builds on the existing "Long Term Evolution" (LTE) standard and constitutes a mobile communications standard that has been gaining ground since 2019.

> of the network that we built as well. So it's a whole encompassing, virtuous thing. [...] So as a testbed, the region itself has all of these different things to offer in terms of those different categories [rural, semi-rural, semi-urban, urban]. And it's I think it's 7700 square kilometres of geographical area that we're looking to cover here in Scotland." (Interview 29, 2020)

The ambitions articulated above indicate the increasingly important role of universities in the co-production of smart cities. For instance, Karvonen *et al.* (2018) emphasise that

> higher education campuses serve as an ideal real-world platform to trial urban innovations because they: (1) comprise a large, single-owner estate; (2) include a collection of buildings and infrastructure networks that are managed in-house; (3) feed into applied research and teaching activities; and (4) enhance the institutional reputation of the university (p. 104).

Crucially, the goal to extend the sensor network beyond the university estate reflects the fact that Edinburgh University makes wider claims related to regional economic development and infrastructure upgrades. These claims were subsequently reiterated in the process of crafting the SIA for the Edinburgh city region.

Following the announcement of regional SIAs in 2015, and building on the findings of the internal University audit in 2014, a local consortium, spearheaded by senior officials from Edinburgh University, started to draft data-driven futures for the Edinburgh city region. Importantly, the envisioning process depicted the attempt to grapple with different geographical scales. Reflecting on the emergence of the idea of the DCoE, key individuals involved described the importance of matching high ambitions with locality, global "imaginaries of innovation" (Pfotenhauer and Jasanoff, 2017a) with regional characteristics. The following statement, made by a senior Edinburgh University representative responsible for industry engagement, sheds light on the brainstorming process:

> "I think when we spoke to Scottish Enterprise, they wanted it to be Scotland as the Data Capital of Europe. We said you can't do that because a capital is a city, it's a place. And so, you know, we went through that process and, you know, to claim that Edinburgh would be the data capital of the world right, would kind of be, that's probably ridiculous and again, that battle is already

> lost because that's in the Valley. [...] So, you know, I mean, so let's go, let's go for capital of Europe." (Interview 25, 2020)

The imaginative work involved in creating the DCoE vision seems emblematic of the multi-scalar nature of place-based innovation. "Data" implies fluidity, mobility and globality. "Capital", in turn, symbolises place, steadiness and permanence. Moreover, "data" appears to constitute an "instrument of imagination" (Beckert, 2016). The possibility to generate, analyse and use big amounts of data holds the promise of imagining distant futures. Big data is imagined as a reservoir of ideas, as signifying constant potentiality. Put differently, and as one official remarked: "data has gravity" (Interview 29, 2020). Setting up a regional IoT network, for instance, promises the university to become a collection point of data, or what Callon (1984) calls an "obligatory passage point". The university is expected to transform into a facilitator for the (re)imagination of the city region by connecting formerly disconnected data sources. Back in the medieval ages, cities were built around rivers and lakes. In the 19th century, canals and railroads shaped people's vision of what a city should be (Cronon, 1991). In the 21st century, cities and city regions are claimed to be built around data or, more aptly, data lochs[13]:

> "You can't be streaming petabytes of data backwards and forwards for the cloud from the UK to America and back again to do your analytics. If you really want to make it work well, locality is still an issue. So making that move is really an important strategic thing for the University of Edinburgh to do, which is why European data capital was coined." (Interview 29, 2020)

3.1.5. So what?

Scales matter when it comes to envisioning desired futures. The DCoE vision reflects complex interlinkages between, within and across historical, political and economic scales. This section demonstrated that urban-regional change, as embodied through the CRD, can be grasped in terms of "the overlapping scales [...] of meaning and action that people deploy to engage with their present, and to assess the consequences of such engagement" (Jiménez, 2005, p. 158). Changing science-society relations and the introduction of policies

---

13 The university's "Data Loch" project constitutes the most obvious example. It seeks to develop a secure repository of health and social care data for the Edinburgh and South-East Scotland region to help find solutions to current health and social care challenges. For more information, see: https://www.ed.ac.uk/usher/dataloch.

such as CRDs, the FCDC and SIAs have opened up (and closed!) spaces for the (re)imagination of the Edinburgh city region (see Figure 20). Another intersection that reveals the non/realignment of actors, organisations, discourses and geographical spaces in the context of Edinburgh's innovation apparatus is the notion of *identity*.

Figure 20: Emergence of CRD | Source: Own illustration

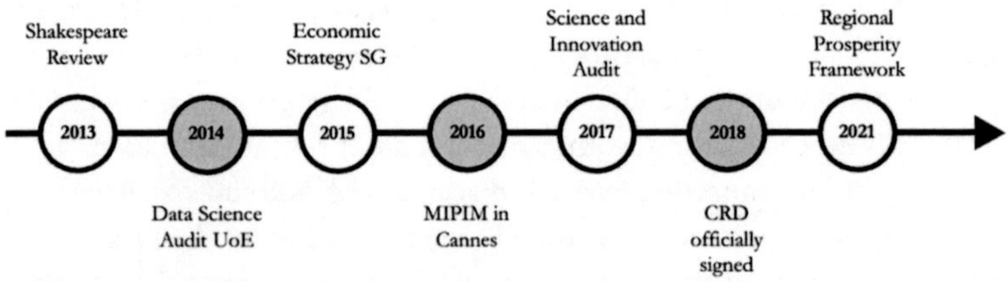

3.2. Identities

Edinburgh's innovation apparatus is also enacted through the interplay of various *identities*. In what follows, I want to illuminate how the CRD and its underlying sociotechnical imaginary articulate complex connections between urban, regional and national identities. More specifically, I am going to show how the promise of transforming the city region into the DCoE is 1) translated into the notion of Edinburgh as "a city of ideas", 2) framed as a potential unifier for the wider region, and 3) performed as part of Scotland's journey as a digitising nation.

3.2.1. Edinburgh as a "city of ideas"

> "It is not that we have the most data scientists or the most data. It's that we are a place that is trusted for its views. So it comes back to this thing about the city as a city of ideas. You know we play to our strengths." (Interview 10, 2020)

The quote above is taken from a conversation with one of the architects of the DCoE vision. It demonstrates that imagining desired futures always implies embedding them into existing ideas, cultures and sociotechnical realities – "the merely imagined is converted into the solidity of identities and the durability of routines and things" (Jasanoff, 2015, p. 323). In Edinburgh's case, that means engaging with a range of historically grown identities. When asked about how they would describe Edinburgh to someone who has never visited the city before, almost all of my interviewees pointed to a few distinctive characteristics: "wealthy", "historic", "successful", "green", "well-educated" and "unequal". (Un)surprisingly, "smart", "innovative" and "data-driven" were not attributes assigned to the city. This begs the following question: How do grandiose and future-oriented imaginaries such as transforming the Edinburgh city region into the DCoE relate to Edinburgh's historical identities?

In the context of the CRD, pasts and futures, history and innovation were linked through the imaginary of Edinburgh as a "city of ideas". Highlighting Edinburgh as a birthplace for great ideas that helped transform the world through scientific progress and creative imagination, data-driven innovation was consciously framed as the logical next step. Invoking the Scottish Enlightenment of the 18th and 19th centuries, one individual, leading on the DDI Programme, reiterates that Edinburgh "has always dealt with concepts, it has always dealt with ideas and governing ideas, management. It is a city of managers in a way" (Interview 7, 2020). One concept that helps shed light on the "management of ideas" is what Farías and Blok call "urban cosmopolitics": the "politics of searching and articulating urban common worlds" (Farías, 2011; Blok and Farias, 2016). Crucially, cosmopolitics "fundamentally involves a process of destabilizing existing propositions of the cosmos and opening a space for the possible and the unknown" (Blok and Farias, 2016, p. 9). Put differently, the embedding of STIs necessitates the (re)arranging, (re)articulation and (de)composing of existing pasts, presents and futures.

Three features of the embedding process are particularly noteworthy. First, key decision-makers, involved in designing the DCoE vision, emphasise the importance of storytelling in place-based innovation projects. One high-ranking official underlines that successful smart city projects exhibit three key characteristics: "a big catalyst programme", "narrators" and "technology" (Interview 10, 2020). In Edinburgh's case, the CRD was used as a vehicle to promote data-driven innovation. Key individuals championing the story included Edinburgh University senior leadership representatives as well as local and national government officials. Finally, the technology element was constantly reproduced by reference to the rich history of data science and AI in Edinburgh. The latter constitutes the second key feature in the embedding process. A fitting example, that highlights the attempt to link history and possibility, are the remarks of a former Vice Principal for Research and Innovation at the University concerning the refurbishment of the School of Informatics:

> "And then there was a very fortuitous fire [Cowgate fire in 2002[14]]. And the building burnt down in the Grassmarket. No, I didn't start the fire, but it allowed us to revisit because one of those departments was lost. We visited and we built a school of informatics, which is, I think, still Europe's largest accumulation of informatitions. And we merged all these separate activities and became really rather a powerhouse and then started translating our fundamental science into industry." (Interview 46, 2020)

Finally, key decision-makers feed off Edinburgh's reputation as a well-educated city when articulating data-driven urban-regional futures. One former leader of the City of Edinburgh Council underscored: "We are a city of smart people. You're always gonna have a vigorous debate in Edinburgh" (Interview 11, 2020). As I will argue later, however, it is important to recognise that "debate" and public deliberation was missing when designing the DCoE vision. Instead, "talent" – one of the five key pillars of the DDI Programme – is predominantly imagined as university graduates working for tech companies or becoming entrepreneurs themselves. This "entrepreneurial understanding" of "smart people" will be picked up in Chapter 6 when I will elaborate on the University of Edinburgh's ambitions to become a "civic university" (Goddard, 2009).

3.2.2. Unifying the city-region?

The CRD and its underlying vision of data-driven progress is also framed as a vehicle for overcoming historically grown tensions within the city region. As mentioned in Chapter 3, the CRD comprises six local authorities: Fife, East Lothian, Midlothian, the Scottish Borders, West Lothian and Edinburgh. The region includes a population of 1.4 million people (24% of the population of Scotland) and contributes approximately £36bn per year to the Scottish and UK economies (City Of Edinburgh Council, 2018). Seeking to illuminate different

14 For more information, see: https://www.scotsman.com/news/day-edinburgh-burned-cowgate-fire-ten-years-1598675

regional identities at play, it makes sense to briefly elaborate on one key policy change: the Local Government Act 1994.

Up until this point, local government was structured around a two-tier system of regions and districts on the mainland and a unitary system in Orkney, Shetland and the Western Isles. With the Local Government Act in 1994, the two-tier system was abolished and the current local government structure of 32 unitary authorities, covering the whole of Scotland, was created. As a result of the reorganisation, for instance, Lothian region council was divided up into four new unitary council areas: East Lothian, the City of Edinburgh, Midlothian and West Lothian.

Tracing the emergence of the CRD, it becomes clear that the local government reorganisation during the mid-1990s heavily affected attempts to create a common identity across the Edinburgh city region, and still shapes current working relationships amongst the six local authorities. One high-ranking council officer at Edinburgh city council remarked:

> "There is a lot of bad blood from that. We are probably just getting over that because it is probably a generation of people that have been involved are no longer involved. So these sort of scars can linger. So I think there has always been an element of distrust. I think you know not all of the partners are equal in terms of size, they all have different challenges, different strengths, different opportunities." (Interview 8, 2020)

According to key decision-makers, the promise of transforming the city region into the DCoE holds the potential to overcome these historical barriers and imagine a common identity. Having said that, one particularly salient issue remains concerning the relationship between Edinburgh and its neighbouring authorities – particularly Mid, East and West Lothian. While officials keep emphasising mutual dependencies, it is important to recognise that Edinburgh – given its economic, cultural and political status – dominates regional future-making.

The rise and fall of "Silicon Glen" substantiates this point. Following the decline of heavy industries such as shipbuilding and mining during the 1960s and 1970s, the Scottish Development Agency (later Scottish Enterprise) promoted electronics manufacturing as a future driver of economic growth. Located in the central belt triangle between Dundee, Inverclyde and Edinburgh, tech companies such as Motorola began producing semiconductors. By the 1980s, Silicon Glen reached its peak and was responsible for manufacturing a third of

Europe's PCs, 80 percent of its workstations and two thirds of its ATMs (Haug, 1986; Fraser, 2016; Masters, 2022). By the early 2000s, the dotcom bubble had burst. What followed was the rapid decline of Silicon Glen with key players laying off employees and closing factories. Nowadays, one interviewee pointed out, areas such as West Lothian have "effectively become an extension of the Edinburgh housing market for people who can't afford to live in Edinburgh" (Interview 6, 2020).

The rise and fall of Silicon Glen illustrates a more general point about regional economic policy. Namely, how to balance centralisation and decentralisation. Up until the creation of the CRD, a coherent economic strategy for the region was missing (Interview 6, 2020). Instead, the region's main development trajectory had been shaped by housing markets. In this sense, the CRD constitutes an attempt to distribute economic growth more equally and promote stronger collaboration between the six local authorities. The partners' recent announcement of a "Regional Growth Framework" (City Of Edinburgh Council, 2019) demonstrates ambitions to articulate a long-term economic strategy for the city region, beyond the lifespan of the CRD. Given the early stage of the CRD, it remains to be seen what lessons will be drawn from the Silicon Glen experience and whether the notion of transforming the city region into the DCoE can actually unify regional partners.

3.2.3. Scotland as a digitising nation

Finally, the CRD and its focus on data-driven innovation must be understood in light of Scotland's ambitions to become a "digital nation" (Scottish Government, 2021a). These ambitions are intimately tied to the Edinburgh city region, the CRD and its main protagonist – the University of Edinburgh. Both, the CRD, through its focus on data-driven innovation, and Edinburgh University as key driver of the DDI Programme, constitute building blocks in driving Scotland's digital economy. Interestingly, data-driven innovation was not always high on the Scottish Government's agenda. One high-ranking official from Scottish Enterprise underlines that "the Scottish Government originally thought data-driven innovation was a waste of money. They didn't understand it. They refused to put it in the city region deal" (Interview 23, 2020).

Only in its *Economic Strategy* (2015) does the Scottish Government introduce "innovation" as an explicit strategic priority. That said, it must be noted that the innovation bit within the CRD is predominantly funded by the UK Government (£290 million) with the Scottish Government "only" contributing £60 million. Since 2015, technological innovations – particularly AI, machine learning and robotics – are at least verbally established as given and massively disrupting technological developments and consciously embedded into the nation's identity as a digitising nation (Scottish Government, 2017, 2021a, 2021c, 2021b).

This becomes obvious when studying the government's latest digital strategy (2021a). One key aim is to grow the digital tech sector by "supporting an innovation ecosystem". This insight goes back to a review of the Scottish technology ecosystem, conducted by Mark Logan, former chief operating officer of Edinburgh-based "unicorn[15]" Skyscanner. The Logan Review (2020) identified "education", "infrastructure" and "funding" as key factors influencing the performance of the innovation ecosystem. Particularly relevant in the context of this book are Logan's remarks on the role of universities. He underlines that "the best universities view industrial collaboration in general, and start-up facilitation in particular, as equally important to their other missions of teaching and research" (Logan, 2020, p. 29). The lead economist, responsible for writing the green book business cases for the DDI Programme, reiterates the need to commercialise innovation by pointing out that "there is a real disconnect between knowledge creation and then the subsequent steps of applying that knowledge creation in innovative ways or through the innovation" (Interview 44, 2020). In Chapter 6, I will pick up this issue by asking what role higher education institutions play in regional innovation economies.

Furthermore, the Scottish Government's appetite to reinvent itself as a "digital nation" (Scottish Government, 2021a) must inevitably be examined against the backdrop of recurring independence efforts. Three examples illustrate this point. First, as mentioned above, the *Scottish Cities Alliance's* smart city programme[16] demonstrates a Scottish-specific approach to smart urban development. Contrary to England, Scotland – because of its size – attempts to speak with one "city voice" when trying to attract investment. Second, the UK Government's commitment to invest into the Glasgow CRD must be viewed in relation to the 2014 Scottish independence referendum. One high-ranking council officer points out that first discussions of a Scottish CRD took place in a highly charged political environment with a case study put together on the eve of the independence referendum: "Scottish government wanted to extract from the UK government maximum monies, maximum influence and UK government wanted to continue to be relevant so there [was] something in it for both of them" (Interview 8, 2020). Finally, key decision-makers seem to

15 The term "unicorn" denotes a company valued at more than $1bn.

16 For more information, see: https://scottishcities.org.uk/smart-cities/

have identified "data ethics" as a potential avenue for a Scottish-specific approach to data-driven innovation. Questioned about Scotland's place in data-driven innovation, one member of the government's Data Delivery Group (DDG) highlights:

> "I think that we have gone through the sort of first wave of data-driven innovation and already collected a fair number globally of bruises and scrapes right, with trying that out. And I think Scotland has the potential to invest in a in a more thoughtful kind of second wave of data-driven innovation that can be that can be piloted here hopefully." (Interview 28, 2020)

Lastly, "Brexit" – the UK's withdrawal from the European Union – significantly shapes Scotland's journey as a digitising nation. There are a few observations that deserve attention. First, the SIA, which paved the way for the CRD, was concluded in November 2016 and published in 2017. Put differently, the decision to position the Edinburgh city region around data-driven innovation and establish it as the DCoE was carried forward even after the Brexit referendum on 23 June 2016. Second, while counter-intuitive and surprising at first, the ambition to become the Data Capital of *Europe* appears to be in line with the Scottish Government's independence efforts as well as the Scottish people's vote to remain within the European Union[17]. Third, while politically committed to the European Union, Brexit is highly likely to have a tangible and harmful impact on the Edinburgh city region. According to the *Regional Prosperity Framework* – the successor of the CRD – "the UK's exit from the EU has the potential to create skills shortages across a range of sectors and occupations, whilst adding to longer-term labour supply challenges related to a shrinking working age population" (The City of Edinburgh Council, 2021, p. 9). This is most evident when considering the significant outmigration of EU nationals from the city region. The number of EU citizens living in the CRD's six local authorities decreased by 37% from December 2019 to June 2020 (The City of Edinburgh Council, 2021). Examining the ambitions of the CRD, Sutherland (2020) succinctly observes:

> "In the light of Brexit the phrase 'of Europe' makes a political statement, albeit rather ambiguously, emphasising links to continental Europe. The 'capital' restates its position as a sub-national capital, albeit with some political

17 For more information, see: https://www.gov.scot/brexit/

aspirations to become a national capital. The question raised is the extent to which the phrase 'Data capital of Europe' has any substantive economic meaning, as distinct from being a marketing slogan to attract businesses and funding." (p. 2)

3.2.4. So what?

An examination of Edinburgh's innovation apparatus is incomplete without taking into account the constant interplay of local, regional, national and global identities. Historical identities such as the notion of Edinburgh as a "city of ideas" shape how decision-makers invoke and imagine data-driven urban-regional futures. Importantly, the embedding of STIs such as the DCoE involves the conscious rearranging of existing pasts, presents and futures. Regional science and innovation projects such as the CRD "encode and reinforce particular conceptions of what a nation stands for" (Jasanoff and Kim, 2009, p. 120). The CRD and its ambition to transform the city region into the DCoE directly feeds into Scotland's ambition to become a "digital nation" (Scottish Government, 2021a). It remains to be seen exactly how the United Kingdom's withdrawal from the European Union will affect the implementation of the CRD. However, there is little doubt that Edinburgh's status as a capital city with a globally radiating and financially dependent university in a potentially independent country will be subject to significant change.

3.3. Agencies

So far, I have predominantly focussed on how the DCoE imaginary was discursively constructed and embedded across different *scales* and *identities*. In this last section, I want to shed light on its *extension* by illuminating human and non-human *agencies*. Jasanoff (2015c) points out that *extension* denotes "the complex of processes by which unconventional ideas gain traction, acquire strength, and cross scales, for example, by persisting through time or by overcoming geopolitical boundaries" (p. 323). In what follows, I am going to showcase three "agents of extension": 1) translation agents, 2) institutions of governance, and 3) physical infrastructures. Importantly, by distinguishing between these three agents of extension, I do not want to suggest an artificial distinction between human and non-human actors. Instead, I fully acknowledge the diverse and distributed nature of agency (Callon, 2008). For the purpose of this chapter, however, and keeping the descriptive focus of assemblage thinking in mind, I emphasise different components of social action.

3.3.1. Translation agents

Key to enacting Edinburgh's innovation apparatus are individuals "who are capable of moving imaginaries from one sociopolitical setting to another" (Jasanoff, 2015c, p. 333). These so-called "translation agents" possess the authority to shape public imagination by defining societal challenges and identifying relevant problem-solving contexts. Michel Callon's (1984) "sociology of translation" provides a helpful starting point for theorising "human agents of extension". As mentioned in Chapter 2, Callon (1984) emphasises that the process of translation consists of 1) defining actors by assigning roles to them, 2) specifying and testing these roles, and 3) rendering actors mobile. Studying scallop cultivation in Saint-Brieuc Bay, Callon illustrates how three researchers become "spokesmen" or "obligatory passage points" for a range of human and non-human actors (e.g., scallops, fishermen, the scientific community): they "obtain the right to express and to represent the many silent actors of the social and natural worlds they have mobilized" (Callon, 1984, p. 224).

Engaging with Edinburgh's innovation apparatus, it becomes evident that one group was particularly influential for the emergence of the CRD and its emphasis on data-driven innovation: the so-called "Gang of Four". Looking up the term "Gang of Four", one stumbles across an 1970's English post-punk band, a political faction composed of four Chinese Communist Party officials and a breakaway group of four Labour politicians who founded the Social Democratic Party in 1981. This particular "Gang of Four" consisted of four senior Edinburgh University officials responsible for injecting data-driven innovation into the CRD: a Senior Vice Principal, a Vice Principal for Research, an Assistant Principal for Industry Engagement and a Vice Principal for Business Development. Asked about characteristics of the group, one member pointed out:

> "We were academically entrepreneurial. We were working in a university. All of us with other skills. And we've all done other things. [...] Again, all of us have come with different sets of capabilities. And I think what we were was ... because none of us were really informatitions, we saw a strength in the university, grasped the opportunity and then were entrepreneurial in leveraging that intellectual capital in order to deliver a step change in the capa-

bility of the university, its surroundings and therefore its region." (Interview 46, 2020)

The significance of this group cannot be underestimated. Acting as representatives of institutionalised expertise, the "Gang of Four" – a name imposed externally – mediated the boundaries between academic capitalism and the innovation economy (Zukin, 2020b), between science and politics. A detailed description of the "Gang of Four" will follow in Chapter 5 in which I am going to illuminate how senior university representatives structure power relationships, and mediate the epistemology-politics intersection in the context of data-driven urban-regionalism.

3.3.2. Institutions of governance

STIs such as the DCoE vision are not only translated through text and talk but also via form and materiality. Governance structures are one particularly important "agent of extension". Once imaginaries get translated into institutional structures, they obtain durability and pave the way for *future* future-making. The CRD governance structure (Figure 21) provides a case in point. Bringing together a range of regional stakeholders (councils, higher education institutions, businesses, third sector organisations) in a complex governance arrangement, it seeks to solidify regional ambitions of sustained economic prosperity. Centred around the Joint Committee, the key decision-making body of the CRD, a variety of sub-bodies are either advising the Joint Committee or are directly delivering the promises made in the CRD. The Joint Committee is comprised of leaders from six local authorities: one university/college sector representative, one business representative, one third sector representative and one representative of the Programme Management Office.

At this point, two empirical insights relating to the make-up of the governance structure are worth mentioning. First, the CRD governance structure seems to reflect a stronger inclination for coalition and consensus in political decision-making. The introduction of UK CRDs demonstrates a willingness to devolve funding and decision-making powers to the city-regional level. Furthermore, in 2007, Scottish council elections moved from the first-past-the-post (FPTP) to the single-transferable-vote system (STV) resulting in increased coalition-building on local government level in Scotland. One former Edinburgh city council leader underscores that because the STV has been in place since 2007, "there's quite a lot of experience of coalition building at a local government level […] which had been a good, positive platform to build on" (Interview 27, 2020).

Figure 21: CRD governance structure | Source: City of Edinburgh Council (2018)

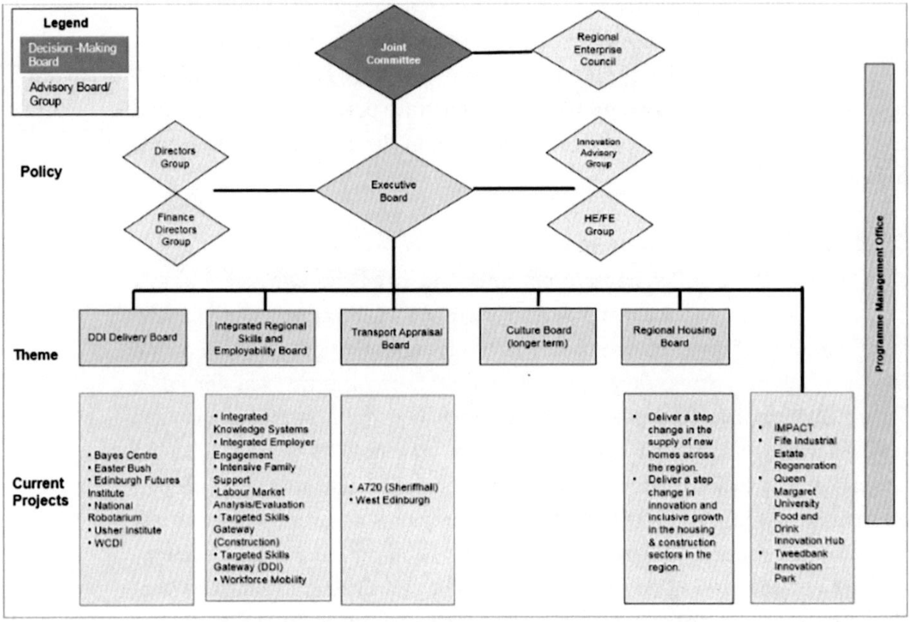

That said, it is evident that the current governance structure does not match the CRD's ambitions around *inclusive* growth. While decision-makers praise the CRD as a vehicle for greater regional collaboration, a coherent and most importantly – inclusive – governance structure seems to be missing. This is most clearly reflected in the creation of the Regional Enterprise Council (REC). Established only *after* the projects and financial arrangements had been agreed in the Heads of Terms and CRD document in 2017 and 2018, the REC is supposed to "provide the voice of the business and third sector to guide the implementation of the City Region Deal" (City Of Edinburgh Council, 2018, p. 34). Having spoken to several members, it becomes clear that the REC – at least during the first two years – did not fulfil more than a token role. One member laments that there were no criteria to judge "inclusive growth" against:

> "When they [Joint Committee] came to us – as I said at the second meeting we rubber stamped 600 million – unless you had that criteria you could have judged it. And also it felt like that they were so far on, what would have happened if we said no? Who would have listened to any changes we suggested? What was our role?" (Interview 5, 2020)

This statement is emblematic of deeper tensions around democratic legitimacy and accountability inherent to the emergence and implementation of the CRD. In Chapter 6, by elaborating on competing imaginaries of the 21st century university, I will address the apparent mismatch between decision-makers' imagined economic futures and local communities' lived realities.

3.3.3. *Building* data-driven futures

Finally, taking into account non-human "agents of extension", constructing Edinburgh's innovation apparatus requires not only social networks and governance frameworks but also material structures and land. In order for STIs such as the DCoE vision to obtain durability, they must be translated into the "practico-material reality" (Lefebvre, 1996) of the city region. There are also more visible materialisations of data-driven innovation in the form of buildings such as the University's five DDI innovation hubs as well as more hidden materialisations like the Advanced Computing Facility or the regional sensor network. In what follows, I will briefly elaborate on the relationship between innovation and architecture by reflecting on two examples that demonstrate how the DCoE vision becomes embedded into urban-regional fabric.

The first example refers to the Bayes Centre (see Figure 22) which is the University's innovation hub for Data Science and Artificial Intelligence.[18] Located in the city centre, it brings together students, researchers, industry and intermediary organisations under one roof. The architecture of the building, particularly its open, flexible workspaces are designed to represent visions of a dynamic, data-driven and interdisciplinary city region. In many ways, it resembles what Sharon Zukin (2020c) calls a "vertical pipeline for tech talent" (p. 185). Taking the Tata Innovation Centre in New York City as an empirical case study, Zukin highlights that different spatial levels signal different roles within the "tech hierarchy". For instance, Cornell Tech's classrooms and studios at the bottom, coworking space for start-ups in the middle and offices for established

18 For more information, see: https://www.ed.ac.uk/bayes

companies on the top floors (Zukin, 2020b). Similarly, the Bayes Centre provides study space, lecture theatres and student cafés on the ground and first floor, with industry collaborators situated on the top levels. In addition, access to the space remains exclusive with students and staff from other schools not being able to enter the building.

But there is also a more hidden, invisible side to the materialisation of the DCoE vision. The Advanced Computing Facility (ACF) of the Edinburgh International Data Facility (EIDF) (Figures 11 and 12) provides a case in point. Built in the 1970s, the facility hosts a range of supercomputers. The UK's national supercomputer ARCHER2 recently replaced ARCHER and possesses 23 racks of servers with 750,000 cores in total. To put this into perspective: a normal laptop possesses four cores (Interview 19, 2020). Located on the outskirts of Edinburgh, it is here where the data-driven innovation infrastructure is created and sustained. In many ways, the ACF depicts the real centre of power since it promises that academia, industry and government will benefit from the collection, use and analysis of large amounts of data.

3.3.4. So what?

Besides *scales* and *identities*, it is human and non-human *agencies* that enact Edinburgh's innovation apparatus. For the purpose of this chapter, I reflected on three relevant "agents of extension" – individuals, governance structures, buildings – responsible for moving desirable imaginaries from one socio-political context to another and embedding the DCoE vision into sociomaterial structures of the city region. Importantly and unsurprisingly, I found support for Davis's and Groves' statement that "the capacity to imagine and enact the urban differently is [...] unequally distributed" (Davis and Groves, 2019, p. 15). Some actors, for instance, the "Gang of Four", possess stronger degrees of projective agency (Emirbayer and Mische, 1998) than others.

One theme running through the rest of this book is the relationship between what can be called the "social stratification of futurity" and subsequent levels of "projective agency". "Social stratification of futurity" denotes the degree to which economic actors are exposed to the escalating nature of capitalist dynamics, and the extent to which their decisions are influenced by futurity. This observation is in line with Pierre Bourdieu's (1979) ethnographic study of the transformation of Algeria during the 1950s, in which he emphasises how members of different social status groups show very dissimilar orientations toward futures. Not only do actors engage differently with futurity but future-imagining practices themselves remain restricted to specific "agents of extension".

4. Conclusion

In this chapter, I have argued that the idea of transforming the city region into the DCoE aligns *actors*, *organisations*, *discourses* and *geographical spaces* around the promise of sustained economic growth and data-driven innovation. Shedding light on three interlinkages – *scales*, *identities* and *agencies* – I have sought to disentangle Edinburgh's innovation apparatus (see Figure 23). To this end, I combined an assemblage-based approach to urban-regional futures (Farías, 2011; Färber, 2014, 2019; Farías and Blok, 2017; Davis and Groves, 2019; Schneider and Lösch, 2019) with Jasanoff's and Kim's (2009, 2015) analytical concept of STIs.

Moving forward, there are three key insights to build upon. First, the emergence of the DCoE vision reflects the multi-scalar nature of place-based innovation initiatives such as the CRD. Three policy changes have paved the way for Edinburgh's innovation apparatus: 1) a renewed appreciation of science and innovation as drivers of societal progress from the mid-2000s onwards, 2) the creation of CRDs in 2011, and 3) the emergence of smart urbanism in 2013. Second, the DCoE imaginary mediates between local, regional, national and global identities. It holds the promise of advancing Scotland's journey as a "digital nation" and "tech nation" (Scottish Government, 2021a, 2022b), overcoming historical tensions within the city region while reinforcing Edinburgh's reputation as a "city of ideas". Finally, Edinburgh's innovation apparatus is enacted and performed through human and non-human "agents of extension". Individuals such as the "Gang of Four" imagine data-driven futures, which subsequently become embedded into governance structures and the sociotechnical fabric of the city region.

This chapter lays the empirical foundation for the remainder of the book. Having introduced Edinburgh's innovation apparatus now three key tensions will be addressed. Chapter 5 will engage with the role of the "Gang of Four" in structuring power relationships and mediating the epistemology-politics intersection in the context of data-driven urban-regionalism. In doing so, it asks: How do issues such as slow growth and inequality become problems of innovation? Why are particular kinds of urban-regional knowledge given importance while other strands of knowledge are side-lined? Why is it that data-driven innovation has been identified as *the* future engine of growth for the Edinburgh city region? How does scientific authority translate into decision-making power?

Chapter 6 will explore the role of powerful higher education institutions such as the University of Edinburgh in enacting the city region's innovation ecosystem. To this end, I ask the following questions: What is the role of global higher education institutions such as Edinburgh University in 21st century urban-regional development? Which competing imaginaries of the modern university are at play when reconceptualising the Edinburgh city region as the DCoE? How do these multiple interpretations work out in practice? At what point are they mobilised and who shapes them?

Figure 22: Bayes Centre Interior | Source: Bennetts Associates

Finally, Chapter 7 provides a meta-level reflection on the relationship between technoscience and planning by mobilising two historical examples: Le Corbusier's high modernism from the 1920s and Salvador Allende's cybernetic socialism from the 1970s. Accordingly, I ask: What role does technoscience play in re-centring the social? That is, aligning social orders around new or different understandings of sociality and the social good. How have "scientific tools" of ordering society changed over time? How does the emergence of big data and artificial intelligence – key ingredients for imagining smart futures in the Edinburgh city region – relate to Le Corbusier's scientific management and Stafford Beer's management cybernetics?

Figure 23: Edinburgh's innovation apparatus | Source: Own illustration

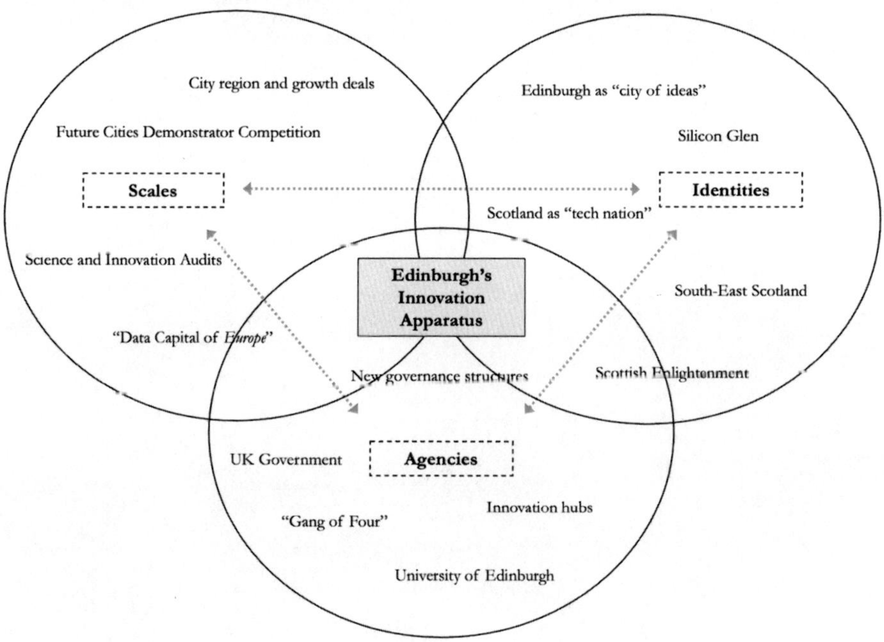

Chapter 5: Solution in Search of a Problem? Data-Driven Innovation and the Politics of Certainty in the Edinburgh City Region

1. A "mad god's dream"?

> But Edinburgh is a mad god's dream,
> Fitful and dark,
> Unseizable in Leith,
> And wildered by the Forth,
> But irresistibly at last
> Cleaving to sombre heights
> Of passionate imagining
> Till stonily,
> From soaring battlements,
> Earth eyes Eternity.
> —Midnight, Hugh MacDiarmid (1921)

These words were written by arguably the most important Scottish writer of the 20th century, and driving force behind the Scottish Renaissance (Riach, 2017). Seeking to reimagine Scotland for the 20th century, MacDiarmid formulated a social and political vision that advocated a "reawakened national purpose and a commitment to egalitarian socialism" (ibid., p. 1). Writing almost exactly 100 years ago, MacDiarmid intended to rediscover the past in order to create a new sense of purpose for the future.

Fast forward to 2016, the lure of a new modernity captured the imagination of a small group of Edinburgh city region leaders. Similar to MacDiarmid, this group attempted to herald the start of a new beginning – a future shaped by data and innovation. In this chapter, I want to ask: Who are the "gods" that dared to reimagine Edinburgh?

The previous chapter laid the empirical foundation of this book by disentangling Edinburgh's innovation apparatus. Shedding light on three interlinkages – scales, identities, agencies – I argued that transforming the city region into the "Data Capital of Europe" (DCoE) aligns actors, organisations, discourses and geographical spaces and functions, as a panacea that promises eternal economic prosperity. This line of argumentation is based on three key insights. First, the emergence of the DCoE vision reflects the multi-scalar nature of place-based innovation initiatives such as the CRD. Second, this particular imaginary mediates between local, regional, national and global identities. Third, Edinburgh's innovation apparatus is enacted and performed through both human and non-human "agents of extension".

Particularly relevant for this chapter's analysis is the first insight. Emphasising the multi-scalar nature of place-based innovation initiatives, in Chapter 4 I looked at three overlapping policy changes which allowed us to reimagine the relationship between science, cities, politics and innovation. Firstly, since the early 2000s, science and innovation in the UK were framed as key drivers of economic growth with the state taking an increasingly interventionist approach. Secondly, the creation of CRDs, from 2011 onwards, has shifted the focus of economic policy from the state to cities and city regions. Finally, since the Future Cities Demonstrator Competition in 2012, cities are increasingly portrayed as sites and drivers of (data-driven) innovation.

In this context, the Edinburgh CRD brings together science, innovation and the city. It embodies changing science-society relations, a strong belief in research and innovation, and an emphasis on cities and city regions as drivers of national economic growth. The co-production (Jasanoff, 2004) of technoscientific and sociopolitical orders opened up an *imaginative space* which reconceptualises city regions as relevant spaces of innovation, higher education institutions (HEIs) as powerful projective agents and science as guiding principle of societal progress (Figure 24).

Having outlined factors which brought Edinburgh's innovation apparatus into being, it is crucial to examine how and by whom this imaginative space was extended and exploited. This chapter puts emphasis on a small group of highly influential higher education officials – the so-called "Gang of Four" (GoF). Studying *human* agency, I want to unpack the coming together of technoscientific and sociopolitical orders via two conceptual lenses from the field of Science and Technology Studies (STS): translation (Callon, 1984; Latour, 2005) and boundary work (Gieryn, 1983, 2018; Mitchell, 1991; Eyal, 2012; Medvetz, 2012a).

Figure 24: Setting the stage for Edinburgh's "architects of imagination" | Source: Own illustration

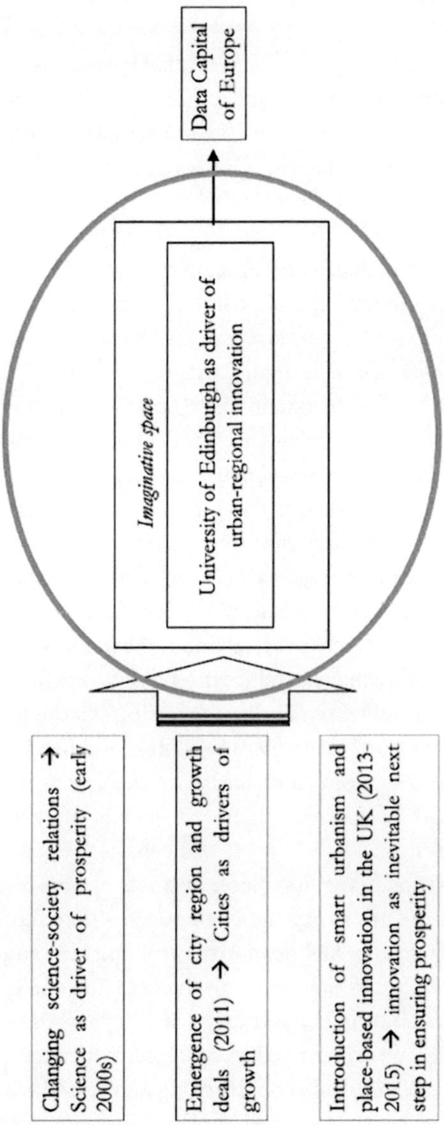

Three core research questions structure the present chapter:

1. How do issues such as slow growth and inequality become problems of innovation?
2. Why are particular kinds of urban-regional knowledge given importance while other strands of knowledge are side-lined? And a related sub-question: Why is it that data-driven innovation has been identified as *the* future engine of growth for the Edinburgh city region?
3. How does scientific authority translate into decision-making power?

In this chapter, I argue that reconceptualising the Edinburgh city region as a relevant space of innovation and positioning the University of Edinburgh as its key projective agent required a process of *apoliticisation* – which can be tentatively defined as the *orchestration of temporal ruptures*. Resting on what Schiølin (2020) calls "future essentialism", scientific actors apoliticise the present by reimagining desirable pasts and producing "an imaginary of a fixed and scripted, indeed, inevitable, future" (Schiølin, 2020, p. 545). In other words, "translation agents" such as the GoF successfully apoliticise innovation policy by mobilising temporal ruptures which result in the side-lining of "old" knowledges and the emergence of "new" facts. As a result, key decision-makers engage in, what can be called, a "politics of certainty", which separates technoscientific progress from political intent (Beck and Wehling, 2013; Gross and McGoey, 2015; Pfotenhauer and Jasanoff, 2017a, 2017b; Pfotenhauer, Juhl and Aarden, 2019). In summary, apoliticisation implies the process of reimagining desirable pasts and linking them to a linear and fixed understanding of the future. Acting as modern "time lords", scientific actors such as the GoF actively engage in the politics of time with the aim of synchronising pasts, presents and futures.

Why apoliticisation in the first place? Two sets of transformations are relevant – one inside the university and one outside – that blur existing lines of distinction between science, politics and the economy, and trigger a rethinking of the political economy of knowledge production. The one inside the university concerns the increasing marketisation of HEIs and the subsequent push to complement the two existing pillars of higher education provision with a third one. Thus the existing pillars of teaching and research are joined by "innovation". The transformation outside of the university relates to the emergence of the city region as a relevant politico-economic space, and the subsequent re-evaluation of HEIs as significant players in urban-regional development policy.

This chapter is divided into four parts. I begin by reviewing existing STS literature on the relationship between science, innovation and the city. Afterwards, I will examine the socioeconomic influence of the University of Edinburgh by elaborating on its relationship with the city region. Section 4 sheds light on the extension of Edinburgh's innovation complex. This will be done by critically introducing the GoF and exploring three modes of apoliticisation: language, location and science.

2. Co-producing science and the city

Dierig *et al.* (2003) remind us that scientific authority and knowledge-producing practices have been involved in the planning, design and maintenance of urban space. Rather than just acting as a resource that shapes urban planning agents' beliefs, science plays a far more complex role in urban environments. The authors discuss four ways in which science and the city intersect: 1) the relationship between scientific expertise and urban politics, 2) science's role in the cultural representation of the city, 3) the embedding of scientific activity in urban socio-material infrastructure, and 4) the interaction between science and everyday urban life. Reflecting on these four intersections demonstrates the co-productionist nature of science and the city: science makes the city and the city makes science.

Crucially, existing accounts in STS, exploring the interplay of science and the city, have neglected the role of HEIs as important agents in urban-regional governance. Noticeable exceptions include scholars such as Andrew Karvonen and James Evans who emphasise the role of "urban experimentation" as a new style of governing cities and city regions (Evans and Karvonen, 2014; Karvonen and van Heur, 2014; Evans, Karvonen and Raven, 2016). For instance, Karvonen *et al.* (2018) shed light on universities as emerging influential actors in the co-production of smart cities. Focusing on the role of university campuses as testbeds for urban futures, they argue that "universities have significant influence on broader economic development, employment opportunities, infrastructure upgrades, and the long-term national and global reputation of the cities in which they are located" (Karvonen, Martin and Evans, 2018, p. 106).

In this chapter, I will build on existing attempts to conceptualise the role of HEIs in urban-regional development by shedding light on scientific actors as central drivers of the evolution and expansion of urban-regional sociotechnical imaginaries (STIs). By shifting the analytical focus away from the state or the corporation as key drivers of urban-regional STIs (Sadowski and Bendor, 2018), I will illuminate how technoscientific and sociopolitical orders co-produce specific representations of the city region. Seeking to understand how science and technology become "enmeshed in performing and producing diverse visions of the collective good" (Jasanoff, 2015a, p. 11), I will zoom in on scientific actors' – senior academics and higher education managers – role in establishing new ways of seeing and knowing the Edinburgh city region.

Aiming to trace the process of apoliticisation, this chapter combines established STS work on "translation" (Callon, 1984; Latour, 1999, 2005) and "boundary work" (Gieryn, 1983, 2018; Mitchell, 1991; Eyal, 2012; Medvetz, 2012a) with more recent STS analyses on innovation policy (Beck and Wehling, 2013; Gross and McGoey, 2015; Pfotenhauer and Jasanoff, 2017a, 2017b; Pfotenhauer, Juhl and Aarden, 2019; Zukin, 2020d). Before delving into empirical detail, I want to briefly revisit the concepts of "translation" and "boundary work" and tease out some theoretical starting points.

How are diverse social worlds aligned in such a way that common representations of knowledge become possible? Key to enacting Edinburgh's innovation apparatus are individuals "who are capable of moving imaginaries from one sociopolitical setting to another" (Jasanoff, 2015, p. 333). These so-called "translation agents" possess the authority to shape public imagination by defining societal challenges and identifying relevant problem-solving contexts. Michel Callon's (1984) "sociology of translation" provides a helpful starting point for theorising "human agents of extension". As mentioned in Chapter 4, Callon (1984) emphasises that the process of translation consists of 1) defining actors by assigning roles to them, 2) specifying and testing these roles and 3) rendering actors mobile. Studying scallop cultivation in Saint-Brieuc Bay, Callon illustrates how three researchers become "spokesmen" or "obligatory passage points" for a range of human and non-human actors (scallops, fishermen, and the scientific community): they "obtain the right to express and to represent the many silent actors of the social and natural worlds they have mobilized" (Callon, 1984, p. 224).

The notion of translation necessarily includes what Thomas Gieryn (1983) coins "boundary work". Gieryn (1995) states that "boundary-work occurs as people contend for, legitimate, or challenge the cognitive authority of science – and the credibility, prestige, power, and material resources that attend such a privileged position" (p. 405). Rather than studying science as demarcated terrain and practice, sociologists should focus on the boundary-work involved in creating the appearance of science as a demarcated terrain and practice. Emphasising the social construction of science, Gieryn (1995) reminds us that actors fill the empty space that is "science" with varied and selectively deployed characteristics, producing cultural maps for justification.

The phenomenon of the boundary problem has been further explored in various other contexts. Most prominently, Star and Griesemer (1989) introduced the concept of the "boundary object". Examining the development of Berkeley's Museum of Vertebrate Zoology in the early 20th century, Star and Griesemer illuminate the importance of boundary objects such as maps, specimen or field notes "which are both plastic enough to adapt to local needs and the constraints of the several parties employing them, yet robust enough to maintain a common identity across sites" (1989, p. 393). Building on Callon's sociology of translation, Star and Griesemer stress that the process of translation across diverse groups (professional biologists, university administrators, amateur conservationists) can be mapped out ecologically without prioritising individual actors/groups over others. The focus is on moments of translation rather than one particular passage point.

A further refinement of Gieryn's "boundary work", that seems useful in the context of this chapter, is what Gil Eyal (2012) calls "space between fields". Seeking to bring Bourdieusian field theory into conversation with ANT, Eyal suggests dedicating boundary work its own space rather than locating it within neither inside nor outside the fields of study (e.g., of science and politics). Instead, he puts forward the notion of the "boundary zone", a thick "zone of essential connections and transactions between [fields]" (Eyal, 2012, p. 162). Particularly helpful is his critique of Gieryn's original conceptualisation of boundary work. In addition to underlining that boundary work must be placed within a separate space, Eyal aptly stresses that boundary work is not limited to rhetoric alone. Instead, it takes robust institutional structures to "limit the number of authoritative speakers, […] to assign their statements with differential values [and to] close off certain topics and devices from nonexpert inspection" (2012, p. 175). In his work on think tanks in America, Thomas Medvetz (2012a) builds on Eyal's notion by characterising think tanks as members of an "interstitial field, or a semi-structured network of organisations that traverses, links, and overlaps the more established spheres of academic, political, business, and media production" (Medvetz, 2012b, p. 25).

One of the aims of this chapter is to explore the boundary work involved in reconceptualising the Edinburgh city region as a space of innovation. To this end, I want to focus on different modes of apoliticisation, which help grasp how scientific actors such as the GoF translate regional sociopolitical orders into technoscientific futures. Importantly, it is not enough to limit an analysis to the GoF's engagement with external actors, such as the public, local and national governments, or the regional start-up scene. It is equally important to examine internal boundary work. That is, what forms of expertise were and are excluded and who decides what evidence is actually used? How is evidence mobilised in urban-regional future-making (beyond mere rhetoric)? It follows that this chapter's analytical focus is less on the formation of specific policies in one field being applied to another. Rather, it is about teasing out the producing and reproducing of lines of distinction, which create the "effect" (Mitchell, 1991) of science as an entity set apart from politics. Adopting this analytical standpoint indicates that problems are not pre-given. In order to "apply" scientific knowledge to urban-regional issues such as inequality, child poverty or slow growth, problems need to be 1) constructed and made visible, 2) framed within a common language, and 3) made amenable to intervention (Miller and Rose, 2008, p. 15).

3. The city and the university

The University of Edinburgh is an urban institution. In order to understand how science and the city co-produce each other, it is crucial to examine the relationship between the city of Edinburgh and its premier university. Founded as "Tounis College" by the town council on 10 October 1583, the university constitutes the oldest HEI in the English-speaking world to be established on a *civic* foundation (The University of Edinburgh, 2020). Unlike other ancient universities in the UK, which were established through the church, Edinburgh University was founded by local people through the town council.

Seeking to provide a brief glimpse into the co-evolution of the city and the university, two figures stand out: Robert Reid (d. 1558) and James Lawson (1538–1584). In his will of 1558, Bishop Robert Reid bequeathed funds to the town council to found a college of higher learning in Edinburgh. Acknowledging Edinburgh's role as the centre of administrative, religious, and legal power in Scotland, and responding to growing calls for an institution to train future ministers, lawyers and state functionaries, the town council, subsequently, petitioned Mary, Queen of Scots, to grant them the site of the former Collegiate Church of St Mary in the Fields "to build a school" (The University of Edinburgh,

2021b). Following years of civil unrest, which included Mary's abdication in 1568 and opposition from the Bishops of Aberdeen, Glasgow, and St Andrews, plans to establish a university were eventually revived in the late 1570s when the town council set up a committee in 1579 to investigate the possibility of establishing a school.

A second figure that was crucial in the foundation of the university was James Lawson. Succeeding John Knox, leader of Scottish Reformation, as chief Minister of Edinburgh in 1572, Lawson became the main advocate of the scheme to erect a university in Edinburgh. At his recommendation, the town council turned to Robert Rollock (1555–1599) – Regent in Philosophy at the University of St Andrews – who became the first Regent and first Principal of the University of Edinburgh (The University of Edinburgh, 2021b). In the centuries that followed, the University developed from a small civic-run college into a modern HEI (Anderson, Lynch and Phillipson, 2003). During the 18th century, under the guidance of Principal William Robertson and Lord Provost George Drummond, Tounis College developed into an enlightened university. Phillipson (2003) points out that "it was the unique interplay between city and college, between a sense of the practical and the philosophical, between learning and public life in a city preoccupied with its own future that shaped the educational culture of the college in the 'Long Eighteenth Century'" (p. 101).

During the 19th and 20th century, the university continued to undergo a complex process of reform and adaptation. Most notably, the Universities Act in 1858[1] took control of the university out of the hands of the city council and placed it in the University Court – up until today, the university's central governing body and legal persona. No longer governed academically, financially and administratively by the town council, the University Court is comprised of the University's Rector, who is elected by the students, the Principal, the Lord Provost of the City of Edinburgh, as well as a range of other academic and non-academic members. Since the Second World War, the university continued to expand in terms of student numbers, academic curriculum and geographical spaces, and is nowadays considered Scotland's premier university (Anderson, 2003b). Ultimately, one cannot imagine the university without the city and vice versa. Founded by the city council in 1583, the University of Edinburgh – to this day – shapes the city and is in turn shaped by the city.

1 More information can be found here: https://www.legislation.gov.uk/ukpga/Vict/21-22/83/crossheading/edinburgh/enacted

Following this brief historical detour, I now want to turn to the 21st century. Before evaluating the meaning and actions of the GoF, it is essential to underline why the University of Edinburgh matters as an urban-regional stakeholder in the first place. In what follows, I will examine key economic indicators of the relationship between Edinburgh University and the city region.

3.1. Economic footprint

Edinburgh's economy is dominated by two sectors, which combined comprise almost half of the total Gross value added (GVA): the finance and insurance sector, which makes up 24% of Edinburgh's GVA, and the public administration, education and health sector, adding £5.3m, comprising 21% of total GVA (see Figure 25).

According to the Edinburgh City Council, 33,000 people in Edinburgh are employed in the education sector, making up 9.6% of all jobs in the city (The City of Edinburgh Council, 2020c). While the accommodation, food and services as well as the finance and insurance industry exhibit similar employee numbers, there is only one industry – health – which employs more people (51,000; 15% of all jobs in Edinburgh). These insights are confirmed when studying the largest employers of the city region. The three largest employers are NHS Lothian (2020), employing 21,921 people, the University of Edinburgh, employing 15,110 people (Higher Education Statistics Agency[2], 2021) and the City of Edinburgh Council (2021), employing 14,898 people. Interestingly, the total staff number (academic and non-academic staff) at the university increased from 9,705 in 2014/2015 to 15,110 in 2019/2020 (HESA, 2021a). According to an economic impact study, carried out by BiGGAR Economics (2015), the university generates £2bn annually for the Scottish economy and supports more than 30,000 jobs across the country.

In addition, it is useful to take student numbers into consideration. Out of a total of 64,025 students enrolled in Edinburgh HEIs, Edinburgh University comprises 35,375 on-campus students and a further 9,135 online distance learners (The University of Edinburgh, 2020). The influence of the University of Edinburgh is further explicated when looking at land ownership in the city region. The University is now the largest landowner of the city,[3] managing 608

2 Forthcoming "HESA".

3 This information was gained in interview research conducted with more than 50 individuals involved in the CRD as part of my doctoral study.

active buildings, and is projected to spend £1.2bn on estate development over the next 10 years (The University of Edinburgh, 2020).

According to the Estates Vision 2017–2027, the university possesses a Gross internal area (GIA) (total area of buildings owned, occupied or maintained) of 887,032 m^2, valued at £1.3bn. Stretching across five campuses, the estate has been increasing by an average of three per cent annually over the last 10 years (The University of Edinburgh Estates Department, 2017). Most recently, the University acquired Edinburgh's Old Royal Infirmary, a Category A listed building at the heart of the city centre, which will host the University's new Futures Institute and is expected to open in 2023. To put these numbers into perspective, there is only one HEI in the UK – the University of Manchester – exhibiting a higher GIA (933,791 m^2) than Edinburgh University. The Universities of Cambridge and Oxford are ranked third and sixth with respective GIAs of 743,873 m^2 and 643,857 m^2 (HESA, 2020).

Figure 25: Gross value added (GVA) in Edinburgh 2018 | Source: The City of Edinburgh Council (2020c)

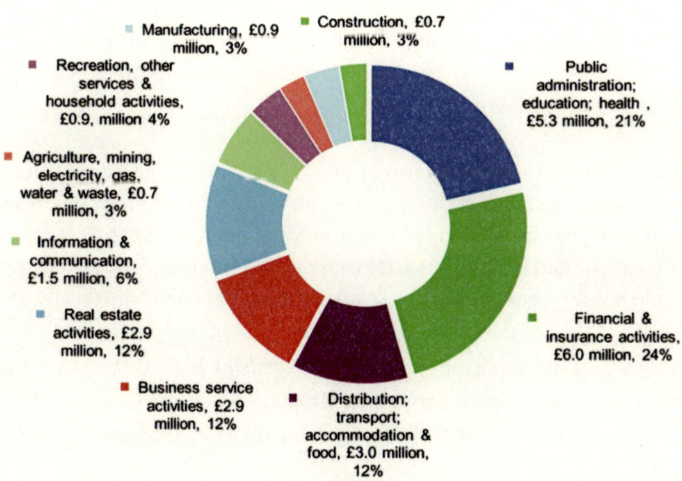

These numbers reveal the university's wide-ranging influence over urban-regional life. Rather than discussing global HEIs such as Edinburgh University in purely educational terms, they must be recognised for what they are: major politico-economic forces (Perry and Wiewel, 2005; Baldwin, 2021). Baldwin (2021) poignantly reminds us that "college and universities have become city managers that, along with producing educational services, also discuss students as consumers, see alumni as shareholders, and imagined the world beyond the campus walls as either prime real estate or a dangerous threat to the brand" (p. 13). Constituting the second largest employer of the city region, the biggest landowner of the city and generating billions for the Scottish economy, the University of Edinburgh represents a key agent in the politics of urban-regional future-making.

3.2. The CRD and the University of Edinburgh

In addition to highlighting the role of the university within the urban-regional economy, it is crucial to briefly reiterate the relationship between the CRD and the University of Edinburgh. As mentioned in Chapter 3, compared to other city region and growth deals in the UK, the CRD stands out due to the significant level of involvement of HEIs. More than half of the Deal's total budget (£791m) is directed towards research, development and innovation. The University's DDI Programme alone receives £350m (The City of Edinburgh Council, 2018), which highlights the UK and Scottish government's commitment to HEIs as significant drivers of innovation. One high-ranking Scottish government official remarked:

> "So I think that's probably one of the most significant characteristics of having one stakeholder in the shape of the University of Edinburgh receiving, I was going to say a disproportionate amount of the funding. That sounds like a criticism and it's not but a lot more of the funding than any other single stakeholder would be getting in any of the other deals" (Interview 47, 2020)

One high-ranking representative of the Scottish Cities Alliance, an advocacy group representing Scotland's seven cities, goes as far as saying that Edinburgh city council is essentially outsourcing its economic development policy to HEIs:

> "The economic development team within the council was much bigger than it is now. But the way that Edinburgh seems to be setting itself up now is very much around that regional model and linking, like basically getting the

universities to lead on its economic development activity through the city region deal." (Interview 32, 2020)

The CRD not only attracted the largest investment of any Scottish CRD but allocates major responsibility, one might even say a "disproportionate amount of funding", to one particular player – the University of Edinburgh – in the hope of reaching national policy goals such as inclusive growth and innovation. Widely regarded as the most powerful HEI in Scotland (Interview 47), Edinburgh University is claimed to serve as a role model for other universities far beyond the Edinburgh city region, interested in engaging more strongly in economic development.

The hopes and expectations assigned to the university are clearly articulated in the CRD *Benefits Realisation Plan* (2020a), a document which sets out to review and evaluate how the CRD is performing against delivery plans, and what economic impacts may be generated as a result. As mentioned in Chapter 3, the University's DDI Programme, which aims to enhance the data capabilities of the city region across ten industry sectors, is structured around five key priorities: talent, research, adoption, data and entrepreneurship ("TRADE"). The CRD's innovation theme, which comprises the DDI Programme, a Food & Drink Innovation Campus and two industrial estate improvement programmes in Fife and the Scottish Borders, is monitored and assessed on the basis of the "TRADE" framework (see Figure 26).

The total expenditure of £791m is projected by the UK and Scottish governments to lead to the involvement of 62,000 students, to deliver a net uplift of £360m of research funding for both Edinburgh University and Heriot-Watt University, to secure an additional £250m of income from industry collaboration and to result in the formation of 440 companies. These projections not only underline the significance of innovation as a key driver of the urban-regional economy but, most importantly, illuminate the promises placed at the University of Edinburgh in leading innovation-based growth for the city region.

Figure 26: TRA(D)E outputs | Source: CRD Benefits Realisation Plan (2020a, p. 31)

Talent	The number of student completions (by programme, course, MOOC[1] or certified CPD[2]) which over the 15 years of the Programme, excluding QMU[3], are projected to involve 62,000 students obtaining formal certification with an additional 30,400 CPDs.
Research	Income awarded to the Universities for the purpose of relevant research and financially managed in each University's research ledger which excluding QMU is currently projected over fifteen years to involve a net uplift of £360 million of research funding.
Adoption	Income from appropriate engagement with individuals/ organisations/industry partners accounted for in each University's finance system. Collectively both Heriot Watt University and the University of Edinburgh anticipate securing an additional £250 million of such funding over the lifetime of the Programme.
Entrepreneurship	Includes the number of companies supported, related funding levels secured as well as students who start up or join DDI related micro/smes. In the former case the anticipated levels are around 440 companies with a minimum (total) level of funding secured of £50 million. In the latter cases around 1000 graduates are projected to be involved.

1 Massive Open Online Course
2 Continuing Professional Development
3 Queen Margaret University

Importantly, the university is not an innocent, selfless enforcer of urban-regional economic policy. Instead, the CRD functions as a vehicle to transform the university from *within*. A look at the university's funding mix illustrates the significance of diversifying income streams. Compared to the two wealthiest UK universities[4] (Cambridge and Oxford universities), Edinburgh University is highly reliant on tuition fees as a source of income. Compared to Oxford (15%) and Cambridge (16%) (HESA, 2021b), tuition fees and education contracts comprised 37% of the University of Edinburgh's total net income (largest source of income) in the academic year 2020/2021 (The University of Edinburgh, 2021a). Furthermore, the University of Edinburgh relies heavily on non-EU fees. Out of a total income of £435m in tuition fees, £272.5m derive from non-EU students.

In light of the Covid-19 pandemic and its impact on international student recruitment, the CRD serves as a vehicle to diversify the university's income streams and push commercialisation activities (Interview 16, 2020). Examples include increased industry collaboration, as reflected in the number of patents

4 Wealth calculated based on total endowment. According to income data by the Higher Education Statistics Agency (2021b) the three wealthiest universities in the UK are the universities of Cambridge, Oxford and Edinburgh.

granted, the number of start-up companies created (114) and the income generated from industrial and transitional awards (£79m) (The University of Edinburgh, 2021a). Additionally, the university received £52m in capital funding via the CRD, which is money to be spent on specific CRD projects. Finally, the CRD and particularly the DDI Programme is claimed to allow the university to build up a local-regional talent pool, thereby becoming less dependent on international labour markets (The University of Edinburgh, 2020).

Having elaborated on the socioeconomic influence of the university by reflecting on its relationship with the city region, I now want to go one step back and ask: How was it possible to frame issues such as slow growth and inequality as problems of innovation in the first place? How did the University of Edinburgh emerge as key driver of data-driven economic growth? To this end, the next section will focus on the GoF, and examine their role in bringing into being Edinburgh's innovation apparatus.

4. Bringing Edinburgh's innovation apparatus into being

In the previous chapter, I argued that three overlapping policy changes in the UK – a renewed recognition of science and innovation as drivers of societal progress, the emergence of CRDs and the development of smart urbanism – reimagined the Edinburgh city region as a space of innovation. In what follows, I want to shed light on how one particularly relevant group of actors – the so-called "Gang of Four" – exploited and extended this imaginative space. I will first introduce the GoF in more detail. Afterwards, I am going to take the Science and Innovation Audit (SIA) as an exemplar for a "co-productionist moment" (Jasanoff, 2004) that manifested the fusion of technoscientific and sociopolitical orders. Finally, I will explore three "modes of apoliticisation" which build the foundation of what can be called a "politics of certainty".

4.1. The Gang of Four

> "And so I have seen when I was in a pub five, six years ago, I got this job and I went to the big property jamboree in the South of France called MIPIM and there was [name] who had just come up from the South of England, where he was chief executive of one of the councils there. And [name] and I flew out together. [...] So anyway, we had a beer and he said, 'Look, I have just done a city deal down in the South Coast. Why don't we do a city deal here?' And

> I said, 'Absolutely. You know, I've just got this job. I've got this big portfolio. Okay, let's ... how do we do this?' He says, 'Look I will write to UK and Scottish government and I will get my foot in the door'." (Interview 21, 2020)

This quote describes an encounter between a high-ranking Edinburgh University official and one of the senior leaders of Edinburgh City Council in 2016. Meeting in the context of MIPIM[5], the world's biggest property market event, these two individuals decided to embark on a journey to bring a city region and growth deal to the Edinburgh city region. In many ways this encounter is emblematic of the strong interconnection between science and politics, the significance of translation agents and the radical de-autonomisation of universities, reflected in the increasing importance of HEIs in shaping regional economic development.

After this first encounter in the South of France, a local consortium or what Kitchin *et al.* (2017) call a "smart city epistemic community" was formed. That is, "a network of knowledge and policy experts that share a worldview and a common set of normative beliefs, values and practices with respect to an issue and help decision-makers identify and deploy solutions to solve problems." (Kitchin *et al.*, 2017, p. 3). This group of urban technocrats consisted of local authority leaders and managers, economic consultants, change management civil servants and higher education officials (both academics and managers). Building on the UK government's *Shakespeare Review*[6] (2013), which concluded that a second wave of economic and societal value will be generated by the "capacity to process and learn from data" (p. 5), the group performed a SIA for the Edinburgh city region, seeking to map research and innovation strengths and identify future areas of economic growth. Subsequently, the SIA formed the basis for the University's DDI Programme.

At the heart of this urban-regional epistemic community were four senior Edinburgh University officials, the so-called "Gang of Four": a Senior Vice Principal, a Vice Principal for Research, an Assistant Principal for Industry Engagement and a Vice Principal for Business Development. As the titles of these individuals suggest, there was a clear division of labour. Two of them (Senior Vice

5 For more information see: https://www.mipim.com

6 The *Shakespeare Review* constituted an independent review that explored the growth opportunities of information held by the public sector. More information here: https://www.gov.uk/government/publications/shakespeare-review-of-public-sector-information

Principal, Vice Principal for Research) assumed "front-of-house roles" (i.e., supervising research and training). They were recruited from the academic workforce. The other two (Assistant Principal for Industry Engagement, Vice Principal for Business Development) occupied "financial and organisational roles", having gained experience in different industries before and risen via administrative roles within universities.

The makeup of the GoF is not unusual. Existing critical literature in the sociology of higher education (Blackmore and Sawers, 2015; Connell, 2019; Münch, 2020) points to the increased fusing of *male* academic and management identities in contemporary senior leaderships teams and even characterises groups, such as the GoF, as denoting a "professional-administrative class" (Schwartz, 2014, p. 515). In Edinburgh's case, the growing influence of managerial identities in higher education leadership is reflected in the emergence of an "entrepreneurial culture", which materialised itself both in the self-perception of the actors and the style of working involved. One senior programme manager, who was seconded from the City of Edinburgh Council to the University and heavily contributed to the SIA, described the GoF as follows:

> "[name] is what I would call an organisational entrepreneur. He's not quite an entrepreneur. But he will make business happen. So he's a deal maker. [name] is someone who creates companies and value. [name] understood the politics. [name] understood the research, how the academic community could be stimulated." (Interview 10, 2020)

This statement is telling not only because it highlights the interlocking of different skill sets (business, politics, research), but also because it reveals the importance of navigating and translating between various audiences. One member of the GoF, who is claimed to have been particularly influential, was consistently praised for his negotiation skills:

> "[name] with phenomenal skills in politics, in dealing with the complexities. There is a lovely picture there of [name] between Theresa May and Nicola Sturgeon. And getting the two of them to sit even roughly on the same platform was very difficult because they hated each other." (Interview 46, 2020)

> "So if there was any hick up, [name] would just get on the phone and he could speak to, he could speak to the first minister, if he wanted to.

I think [name] was such a diplomat that what he did was he could see movement." (Interview 44, 2020)

"Dealing with complexities" and "seeing movement" appear to have been crucial traits in order for the GoF to engage in successful boundary work and establish data-driven innovation as an "organising idea" (McQuillan, 2018). The emergence of the DDI Programme provides a case in point. Following the publication of the SIA in 2017, which identified DDI as *the* future engine of growth for the Edinburgh city region, the UK government initially did not provide money to fund the programme. Searching for alternative funding streams, the GoF identified the CRD as the ideal vehicle to carry forward its ambitions around DDI. One member of the GoF remembers this moment as follows:

> "So when we actually published it, you know, we said to government, 'So how are we going to get this implemented? You know, there is momentum to do this.' And they said, 'There's no money.' So at that point, it essentially became, it went into the city region deal as one of the options for consultation with government. And actually at that point, that would be Christmas, Christmas/New Year '17 ... well '17, '18 ... government came back and said, 'Of all the ideas you've put forward, the only one that's really interesting is DDI. Just work on that'." (Interview 25, 2020)

The "academically entrepreneurial spirit" (Interview 46, 2020) was further enhanced by what members of the GoF called an informal, "start-up-like" working culture. The following two quotes by this individual are illustrative of this point:

> "So we were meeting once or twice a week and very informal. No notes of any no meetings really, etc. We came together. We had a white board. [...] we just came together, we just drew stuff." (Interview 21, 2020)

> "So when we presented the data driven innovation programme, I saw it as a start-up company, as a start-up programme that should have its own governance, that should have its own board and it should report up through the university hierarchy." (Interview 21, 2020)

Asked about future university governance, the same individual goes on to say

> "I mean, I really, really would be looking at seriously restructuring the university. I would look at the bottom 20 percent of academics and I would get rid of it. I would rationalise the administration significantly. I would build a concept of one university." (Interview 21, 2020)

The last quote is particularly striking since it demonstrates that the ambitions of the GoF were not restricted to the DDI Programme but also aimed at reimagining and transforming the university from *within*. This observation is reflected in the composition of the GoF, which reveals the growing influence of professional, non-academic administrators. The individual cited above constitutes a case in point. Guided by financial imperatives such as student numbers and staff cost, they embody what Lynch *et al.* (2012) call "new managerialism" – a management strategy that "involves governing through enacting technical changes imbued with market values" (Lynch, 2015, p. 193).

More generally, the fusion of economic and scientific practices – most notably the increased focus on innovation and entrepreneurship in institutional strategies and the growing importance of non-academic staff in university leadership – reflect what social scientists characterise as the rise of "academic capitalism" (Slaughter and Rhoades, 2004; Tuchman, 2009; Münch, 2020). That is, "a unique hybrid that unites the scientific search for truth and the economic maximization of profits" (Münch, 2020, p. 1). Examples include "new circuits of knowledge", which link scientific knowledge production directly to the production of marketable goods (e.g., patents), the emergence of "interstitial organisations" such as economic development offices or "enhanced managerial capacity", which aims at improving external competitiveness and internal controlling processes (Slaughter and Rhoades, 2004). The GoF and its efforts to capitalise on the city region as an imaginative space of innovation, by transforming the university from within, seem emblematic of what can be called "academic capitalism for the innovation economy" (Zukin, 2020b).

4.2. Constructing data-driven futures

Before engaging with three modes of apoliticisation, it makes sense to briefly reiterate the significance of the SIA, which was completed in 2017. In what follows, I want to characterise the emergence of the SIA as a "co-productionist" moment (Jasanoff, 2004). Jasanoff (2004) reminds us that scientific knowledge "embeds and is embedded in social practices, identities, norms, conventions, discourses, instruments and institutions" (p. 3). In Edinburgh's case, technoscientific and sociopolitical actors co-produced a specific representation of the city region: the Edinburgh city region was imagined as the DCoE.

Conducted by the GoF and a range of local authority leaders and change management civil servants, the SIA laid out an economic vision for the Edinburgh city region that brought together scientific authority and knowledge-producing practices on the one hand and socioeconomic power on the other. Importantly, it resembled what Schiølin (2020) calls "future essentialism": a vision that "produce[s] and promote[s] an imaginary of a fixed and scripted, indeed inevitable, future, and that can be desirable if harnessed in an appropriate and timely fashion, but is likewise dangerous if humanity fails to grasp its dynamics" (Schiølin, 2020, p. 545). The essentialist nature of the DCoE imaginary can be unpacked via three discourses: the "dialectics of pessimism and optimism", "inevitability", and "epochalism" (Schiølin, 2020).

The first discourse is characterised by its underlying binary logic (see Figure 27). According to the SIA, there are only two options for the Edinburgh city region: either invest into data-driven innovation *now*, and thereby secure economic prosperity, or defer and "run the risk of losing both competitiveness and output to other leading digital clusters that have the confidence to invest" (UK Government, 2017, p. 2). Resulting from the "commit now or defer" logic is a sense of "inevitability". Because there are only two options, acting on the more desirable one (committing to DDI) is inevitable. The inevitability of investing into DDI is most clearly articulated in the SIA's vision:

> "Harness the disruptive potential of data science to create economic gains, unlock value from public data and deliver sustainable gains in private sector activity, public sector transformation and citizen well-being whilst addressing the regional digital skills gap" (UK Government, 2017, p. 65)

Finally, the SIA invokes a discourse of "epochalism" meaning "a sense of tenseness and suspense of living in extraordinary times" (Schiølin, 2020, p. 549). This is reflected in the SIA's persistent reference to what Klaus Schwab, founder of the World Economic Forum (WEF), calls the "fourth industrial revolution" (Schwab, 2016). For instance, the authors repeatedly stress that "data and its manipulation lie at the heart of the next phases of the digital and data revolution" (UK Government, 2017, p. 27).

To sum up, the SIA is riddled with "future essentialism", painting a future that is fixed and one-directional and does not allow for alternative imaginaries. This finding is in line with what Pfotenhauer *et al.* (2019) call the "deficit model of innovation". That is, data-driven innovation is "wielded as a political framing device that constructs parts of society and various social institutions as insufficient and in need of an innovation agenda" (p. 896).

Crucially, however, technoeconomic progress is not divorced from political intent. Instead, innovation is "political all the way through" (Pfotenhauer and Juhl, 2017, p. 88). In the context of the CRD, the political nature of innovation is most prominently reflected in the reconfiguration of urban governance structures. The key question, therefore, becomes: How can projective agents such as the GoF drive innovation while maintaining the appearance of apoliticism?

Figure 27: Dialectics of data-driven innovation | Source: Science and Innovation Audit (UK Government, 2017)

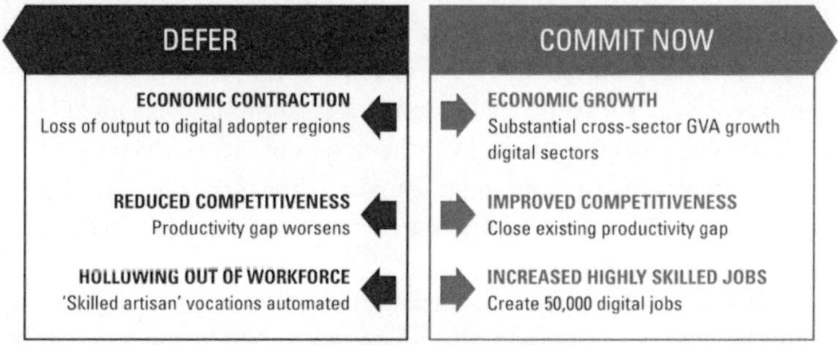

4.3. Three modes of apoliticisation

Three prevalent *modes of apoliticisation* enabled the GoF to advance its innovation agenda without overstepping existing lines of distinction: language, location and science. Forming the basis of what I have called the *orchestration of temporal ruptures*, these three modes allow to reimagine desirable pasts and produce an imaginary of a fixed and inevitable future.

4.3.1. Innovation talk

> "It might sound slightly cynical, but you need something like that. You need new phrases. It's like, you know, big data or whatever. You need something that lights people's imagination and pulls them in." (Interview 25, 2020)

Language matters. The quote above highlights that a lot of the "imaginative work" involved in creating the DCoE vision was dedicated to finding "the right words". Miller and Rose (1990) aptly emphasise that language constitutes a mechanism or "'intellectual technolog[y]' that render[s] aspects of existence amenable to inscription and calculation" (p. 1). Put differently, in order for Edinburgh's innovation apparatus to become governable, actors such as the GoF must inject "vocabularies of innovation" (see Figure 28) that key decision-makers abide by. Engaging with the GoF's efforts to establish a common language, two observations stand out: first, the importance of narrators championing the story and, second, the interpretive flexibility of key terminology.

Unsurprisingly, unless key decision-makers buy into imaginaries of innovation, any innovation project stops dead in its tracks. For data-driven innovation to materialise "on the ground", it had to be established discursively first. The phrases "data-driven innovation" and "Data Capital of Europe" serve as prime examples for the elasticity of language, allowing actors to imagine a "landscape of innovation where everyone prospers" (Zukin, 2020a, p. viii). The emergence and subsequent adoption of these terms is worth highlighting. According to one member of the GoF, the notion of "data-driven innovation" was heavily driven by Jon Oberlander, founding director of the Bayes Centre – the University's "innovation hub" for data science and artificial intelligence:

> "So Jon Oberlander and I came up with this, I think it was actually Jon who came up with this term 'data-driven innovation'. And so now you kind of go out and everyone talks about data-driven innovation. But actually, at the time, no one was, no one was really doing that. He kind of either picked that phrase up or coined it. And so we kind of adopted it." (Interview 25, 2020)

Figure 28: Commonly used "vocabularies of innovation" | Source: Own illustration based on interview data

"World class"	"Doing data right"
"Data-driven innovation"	"Data revolution"
"Data has gravity"	"World-leading research"
"Data Capital of Europe"	"Unlock potential"
"Ubiquitous and disruptive nature of DDI"	"Knowledge economy"

Apart from the obvious hubris surrounding this statement, it is crucial to understand the power of "buzzwords" such as "data-driven innovation". Observing their performative nature, Rist (2007) rightly stresses two key elements: "an absence of real definition, and a strong belief in what the notion is supposed to bring about" (p. 486). Initially described as the "use [of] data to shape, develop and deliver to consumers and citizens innovative processes, including new digital products and services" (UK Government, 2017, p. 4), data-driven innovation has been established as a panacea for solving complex social issues in the Edinburgh city region. Firmly rooted in the grand narrative of the Enlightenment, key decision-makers such as the First Minister of Scotland, Nicola Sturgeon, or the university's Vice-Chancellor, Peter Mathieson, invoke "Edinburgh's data revolution"[7] and the opportunities linked to transforming the city region into the DCoE.

The latter constitutes a particularly interesting case in demonstrating the interpretive flexibility of innovation talk. Asked about the meaning of this particular imaginary, it becomes clear that the notion of the DCoE predominantly acts as an empty signifier or "boundary object" (Star and Griesemer, 1989), allowing multiple social domains to communicate with each other. Importantly, there was no common definition of DCoE. Some respondents labelled the imaginary as a "handy headline" (Interview 48, 2020), which signals the city region's ambition to transition into a modern knowledge economy. Others admitted not knowing much about "data", let alone "data-driven innovation", but stressed their confidence in the university's capability to steer future economic growth. And yet others – although a small minority – expressed significant doubts:

> "I have absolutely no idea what that means Nicolas, absolutely no idea. Absolutely not. […] You know, I understand its [data's] value but I wonder what

7 This phrase was recently used by Peter Mathieson, Vice-Chancellor and Principal of the University of Edinburgh, at the Scotsman Data Conference 2020.

> being the Data Capital is. To me it's just like, maybe I don't know enough about it, maybe I am too stupid, maybe I have a problem with attention or maybe it's just one of those things that people say that sound awfully good. I told you I was quite cynical." (Interview 40, 2020)

Crucially, since resistance remained marginal (or at least couldn't gain sufficient momentum) and key stakeholders championed the idea of data-driven innovation, the imaginary of the DCoE established itself as a "container" to advance data-driven futures. It was precisely this vagueness that allowed it to serve as a "boundary term" (Scoones, 2007) linking diverse social groups (politics, science, business) with diverse interests. Most strikingly, the discourse around data-driven innovation was conducted in an *apolitical* manner. The supposedly "trans-ideological properties" (Fox, 2007) of innovation were elevated and cumulated in, what critical innovation scholars call, a "pro-innovation bias" (Godin and Vinck, 2017). In other words, key decision-makers were guided by the assumption that technological innovations are inherently attractive and inevitably portray the "next big thing" (Interview 7, 2020). The following statement by a senior City of Edinburgh Council member demonstrates the naturalisation of innovation:

> "I don't think 'Data Capital of Europe' is necessarily that thing but it is a useful thing to project to the world in terms of what Edinburgh actually is. Because we are more by embracing new technology and being a centre of innovation." (Interview 1, 2019)

4.3.2. Edinburgh as a "truth-spot"

Performing the DCoE vision goes beyond the discursive domain. It also materialises in *urban-regional space* because the production of knowledge goes hand in hand with the production of space. Gieryn (2018) reminds us that place matters for what people believe to be true. In what follows, I want to apply his concept of "truth-spots" (Gieryn, 2006) to the extension of the DCoE vision in the Edinburgh city region. Doing so, I ask: How does the mobilisation of place contribute to the apoliticisation of the GoF's innovation agenda?

As mentioned in Section 3.1., the university is inextricably linked with the city region. In many ways, extending the DCoE vision across the city region implies "re-writing" (Lefebvre, 1996) past and existing infrastructures, methods, regulations, buildings and ideas. Invoking a Lefebvrian understanding of the urban, it becomes clear that the production of urban-regional innovation space affects both "the city" and "the urban". That is, it shapes both a "present and immediate reality, a practico-material and architectural fact" and "a social reality made up of relations which are to be conceived of, constructed or reconstructed by thought" (Lefebvre, 1996, p. 103). In what follows, I want to highlight these two dimensions by reference to the six innovation hubs.

Ultimately, the CRD's DDI Programme constitutes an infrastructure project. The overall budget of £791m depicts capital investment into the DDI Programme's six innovation hubs, a Food & Drink Innovation Campus and two industrial estate improvement programmes in Fife and the Scottish Borders. Setting up Edinburgh's innovation apparatus, first and foremost, entails *building* physical infrastructure. The six DDI hubs – the Bayes Centre, the Edinburgh Futures Institute, Easter Bush, the Usher Institute, the National Robotarium and the Edinburgh International Data Facility (EIDF) – represent "truth-spots" (Gieryn, 2006), which help stabilise the reimagination of the city region as a space of innovation.

Gieryn (2018) reminds us that the Oracle of Delphi – the "Mother of All Truth-Spots" (p. 17) – once constituted a place where people received an authoritative consultation about the future. During the nine warmest months of each year, Pythia, the high priestess of the Temple of Apollo, would give prophecies on everything, ranging from when a farmer should plant their seeds to the possibility of looming war. Edinburgh's six innovation hubs seem to represent oracles of a new form of urban-regional modernity. What used to be God is now data. What used to be Pythia are now scientific entrepreneurs and innovators. One of the six hubs considered to be of particular importance for enacting Edinburgh's innovation apparatus is the EIDF. Hosting the UK's national supercomputer, the EIDF provides the enabling data infrastructure platform for the wider DDI initiative. It is expected to ensure the storage and flow of data.

One DDI project that was repeatedly mentioned by interviewees and was said to portray a direct outcome of the investment into increased computing power is the "Data Loch[8]". It seeks to develop a secure repository of health and social care data in order to meet challenges, such as an ageing population or surging drug costs. As already touched on in the previous chapter, initiatives such as the Data Loch allow the university to position itself as an "obligatory passage point" (Callon, 1984), thereby structuring the ways in which scientific knowledge production and "innovation" unfold. Crucially, the mobilisation of

8 For more information, see: https://dataloch.org/about-us

places such as the EIDF promotes a form of scientificity that derives its credibility from a healthy degree of "data determinism". The New York Times columnist David Brooks accurately grasps the gist of it.

> "We now have the ability to gather huge amounts of data. This ability seems to carry with it certain cultural assumptions — that everything that can be measured should be measured; that data is a transparent and reliable lens that allows us to filter out emotionalism and ideology; that data will help us do remarkable things — like foretell the future." (Brooks, 2013)

It follows that invoking Edinburgh as a "truth-spot" involves the careful orchestration of "the city" *and* "the urban". It is not enough to build innovation infrastructure. It requires translation agents such as the GoF to enrol and mobilise actors and places. The six innovation hubs seem emblematic for what Zukin (2020d) calls the "urbanisation of Silicon Valley" (p. 3). Apart from visually breaking with Edinburgh's predominantly Georgian architecture through the injection of futurist forms and materials (glass, steel, verticality, transparency), the DDI hubs appear to be "charged" with a particular understanding of the social. They are built for a "society of entrepreneurs and service users" (Cowley, Joss and Dayot, 2018). Rather than promoting modalities of publicness, that emphasise "civic" and "political" elements of sociality, they envisage "tech talent" – citizens who consume everyday urban-regional services and create economic value. This particular embedding of scientific activity in urban socio-material infrastructure (Dierig *et al.*, 2003) depersonalises scientific knowledge production by shifting agency from people to data. As a result, data-driven innovation sits comfortably in an apolitical, uncritical and abstracted space.

To summarise, infrastructures such as the six innovation hubs can trigger an "enthusiasm of the imagination" (Mrázek, 2002, p. 161). They set out to remake urban-regional fabric by signalling a new beginning. They deliberately tap into Edinburgh's identity as a "city of ideas" (Interview 10, 2020) framing "data-driven innovation" as the logical next step in technoscientific progress. Reflecting on the emergence of tech space in Edinburgh, one prominent technology columnist at the Scotsman goes as far as terming Edinburgh's tech scene as the city's "Holy Trinity of Tech" (Freer, 2017). The lure of this new modernity must have captured at least someone's imagination.

4.3.3. Mobilising Data Science

Finally, besides language and location, it is science that matters. More specifically, *data* science. In what follows, I want to explore how the GoF mobilises data science not just as a method but as a way of seeing and enacting the social world. In doing so, I argue that data science functions as an "organising idea" (McQuillan, 2018), which allows the GoF to reimagine desirable pasts and advance personal careers. Underlying this mode of apoliticisation is a conviction to see data science as the description of a hidden layer of reality rather than understanding it as part of the production of reality itself.

Pursuing this line of argumentation, it makes sense to briefly revisit Yaron Ezrahi's (1990) remarks on the relationship between science, technology and liberal democracy. Studying four centuries of cultural history, Ezrahi argues that science has been mobilised as a resource for the construction and legitimation of political power. Emphasising the importance of "[scientific] instrumentalism as a structure of accountability", Ezrahi states that "scientific-technical norms have made it possible to substitute technical discipline for moral, organizational and political controls as socially trusted guarantors of the integrity of public actions" (Ezrahi, 1990, pp. 43–44). The result is a sociocultural space in which human action is *externalised* as a relation of causes and effects and political actors appear to be acting for *all* people, thereby upholding horizontal rather than hierarchical accountabilities.

Similar to what Ezrahi describes as "instrumentalism" and the "externalisation of human action", the GoF appears to be speaking with the voice of science in order to legitimate, and, most importantly, apoliticise its own actions. They embrace the scientific pattern of "standing outside", or what Donna Haraway (1988) calls the "view from nowhere", by linking data science to cultural norms of scientific practice such as objectivity, impersonality and representativeness. Crucially, however, there seems to be a fundamental difference between "science" and "data science". Contrary to science, which seeks to distance itself from knowledge implementation in order to keep up the "veil of neutrality" (McQuillan, 2018, p. 262), data science "not only makes possible a certain way of knowing but also acts directly on the knowledge produced" (ibid.). Amongst many other examples, this has become evident in research on the relationship between child welfare investigations and poverty, which showcases the problematic use of algorithms and big data to identify children at risk (Eubanks, 2018).

The uneasy relationship between data-driven innovation and data science comes into the open when confronting higher education managers with the "politicisation of science" and the "scientisation of politics" (Weingart, 1999). Asked if scientists act as policymakers, this individual, slightly irritated, replied as follows:

> "I am not going there. No, that's not a label you could put on this. They're not policy makers. What they're doing is getting the evidence to create policy. You can inform it. You can have a point of view, but you don't set the policy. That is in the realms of politicians." (Interview 34, 2020)

This statement is striking for at least two reasons. First, this actor clearly separates technoscientific progress from political intent. In other words, science – as embodied through the University of Edinburgh – provides evidence-based research, which politics then transforms into policy. Second, data science is not seen as a "social apparatus" (Foucault, 1980) but simply as the latest application to the solution of complex social problems.

Data science, however, is not just mirroring reality. It constitutes a specific representation of reality. A view from nowhere is always a view from somewhere. Applied to the underlying empirical case study, it denotes a specific way of *seeing* the city-region. Separating science from politics and method from apparatus allows actors such as the GoF to construct slick narratives and make powerful knowledge claims. The following statement – taken from the University's annual report 2021 – seems emblematic for the glorification of data-driven research:

> "Data Loch will enable researchers to engage with private, public and third sector health and social care organisations, using data from across the Edinburgh and South East Scotland region. This was followed in February 2021 with the purchase of the world's fastest AI computer, the Cerebras CS-1, for the Edinburgh International Data Facility. EPCC, the University's super-computing service, will use Cerebras CS-1 for unprecedented AI scalability and massive data handling capability, ultimately providing a fantastic facility for DDI activity and research." (The University of Edinburgh, 2021a, p. 1)

Having characterised data science as an "organising idea" (McQuillan, 2018), I want to briefly zoom in on two empirical articulations of the latter: history and career-building. As for the first element, it is noticeable how actors such as the GoF reimagine desirable pasts in order to construct seemingly linear futures. Mobilising history presents a convenient tool to gloss over the mundanity of the CRD. Besides invoking grand Scottish Enlightenment ideals, members of the GoF repeatedly emphasise Edinburgh's rich history of AI and data science:

> "But if you look at the UK basis, it [data-driven innovation] is distinguishing us. There are many reasons. If you go back to the original papers why that's right. One of the birthplaces of AI, '63. We've got worldwide collaborations with Harvard. So many inventions and areas of research have come out of Edinburgh." (Interview 34, 2020)

This narrative is complemented with the persistent invocation of early research collaborations such as the "Edinburgh Stanford Link"[9] (2002), the city region's Silicon Glen experience (1980s) and the university's role as a founding member of the Alan Turing Institute (2015). These examples illustrate that pasts have to be rearranged in order to create desirable futures. The future is always also a *historical* project (McFall and AWED, 2020).

A second empirical observation that underlines the influence of "data science" as a way of seeing the city-region relates to the advancement of individual careers. One senior manager at the Edinburgh Parallel Computing Centre (EPCC) and Professor in High Performance Computing reflects on his rise within the University as follows:

> "So I find it very interesting from my personal career that suddenly I stepped up in the university and I was going to meetings I never expected to go to. So I watched, I was called to a meeting where I was the only person that didn't have 'Principal' in their job title." (Interview 19, 2020)

This individual goes on to describe a meeting between the GoF and the UK government's chief civil servant in 2017, which led up to the integration of DDI into the CRD:

> "And he [civil servant] said to [university official], 'We are very interested in your Science and Innovation Audit. What do you think it would take to deliver it?' And [name] said, 'Well, obviously money is very tight, but we feel some investment will be required.' And the civil servant replied, 'Well, money can always be found for the right projects.' So the civil servant said, 'So what do you think would be the sort of appropriate amount of money?' And [name] said,

9 For more information, see: https://en-academic.com/dic.nsf/enwiki/8189946

> 'Well, we've thought about this and we think around 300 million pounds investment for the region'." (Interview 19, 2020)

When questioned about their role in the negotiations, they pointed out:

> "And I thought, oh, I'm here because they're showing me how you do that. […] Well, I have never asked government for 300 million pounds just like that. So I reached a point in my career where I am being shown how to do that." (Interview 19, 2020)

This statement is remarkable. Not only does it provide insight into high-level negotiations between a powerful university and the UK government. More importantly, it reveals how the GoF engaged in internal boundary work. Prioritising and mobilising specific forms of expertise over others, some actors are granted permission to engage in the politics of future-making while others are left outside. Interestingly, this example is not an exception. At the time of writing, the university promoted the Director of the Bayes Centre to become Deputy Vice Principal of Research with a particular focus on developing the institution's AI strategy. It follows that the mobilisation of data science is not restricted to the CRD but features heavily in the university's strategic direction.

Thus, the GoF mobilise a specific form of scientific expertise – data science – when it comes to engaging in the politics of urban-regional future-making. This observation raises further questions: How does data science, as a form of scientific expertise, intersect with urban-regional politics? (Karvonen *et al.*, 2021) Does data science produce specific ontologies of the city-region? If so, does this way of "seeing the city-region" clash with existing understandings of urban-regional life?

5. Conclusion – a "politics of certainty"?

This chapter was inspired by a desire to understand how complex social issues such as inequality and child poverty become "problems of innovation". I was particularly interested in exploring why it is that certain forms of knowledge gain momentum while others are side-lined. Engaging with these questions, I argued that reconceptualising the Edinburgh city region as a relevant space of innovation and positioning the University of Edinburgh as its key projective agent required a process of apoliticisation. Defining the latter as the *orchestration of temporal ruptures*, I demonstrated how the GoF mobilised language, location and data science in order to construct "an imaginary of a fixed and scripted, indeed, inevitable, future" (Schiølin, 2020, p. 545).

One key insight emerging from this chapter is the recognition that "knowledge games" are inherently ideological and take place within and across institutional settings. Studying the Edinburgh city region, there seem to be at least two transformations taking place which trigger a reconceptualisation of existing science-politics boundaries. First, increasing pressures to commercialise research and establish "innovation" as a core pillar of higher education provision. Second, the (re)emergence of the city region (Martin *et al.*, 2019) as a relevant space of innovation, which results in changing governance structures. Caught between those upheavals, the GoF were confronted with the challenging task of responding to shifting boundaries without giving the impression of becoming a political actor. Deeply rooted in what Schiølin (2020) calls "future essentialism", the GoF engaged in a "politics of certainty" by separating technoscientific progress from political intent.

A range of questions emerging from this chapter invite further exploration. Having shed light on human agency by way of focusing on the actions of a specific group, the next chapter of this book will move up the analytical ladder to examine the university as a key institution in 21st century urban development. I will engage with the following questions: How does the university, as an institution but also as a site of knowledge production, affect necessary transitions of the 21st century? What are the conditions that allow HEIs to move from humble servants of the public good to powerful futuremakers? How does scientific expertise frame our ideas about the future, and hence our ideas about how to make decisions in the present? Finally, can we speak of a "neo-scientisation" of urban development?

Chapter 6: Competing Imaginaries of the 21st Century University

1. Introduction

> "Founded by Edinburgh's Town Council in 1583, we are the oldest university in the English-speaking world to be established on a civic foundation and, as such, place a special importance on how we engage with our city, our region and its various communities." (The University of Edinburgh, 2020, p. 2)

> "[...] this [the CRD] is not a piece of democracy. This is a vision by a relatively small number of people trying to drive up the capabilities of the region." (Interview 46, 2020)

The two quotes above capture a fundamental tension in the relationship between the *university* and the *city region*. On the one hand, the University of Edinburgh imagines itself as a civic institution, emphasising the importance of placing its research and education in the service of the community. On the other hand, its key officials negotiated a backroom deal on the future of the city region with the UK and Scottish governments that was marked by its opaque and exclusive genesis. How does that go together?

In this chapter, I will address this tension by engaging with competing imaginaries of the 21st century university. I ask: What is the role of global HEIs such as Edinburgh University in 21st century urban-regional development? Which competing imaginaries of the modern university are at play when reconceptualising the Edinburgh city region as the "Data Capital of Europe" (DCoE)? How do these multiple interpretations work out in practice? At what point are they mobilised and who shapes them?

Engaging with these research questions, I argue that the University of Edinburgh's shift from traditional self-imaginations such as the "ivory tower" to more globally, networked imaginaries, embodied in the notion of the "entrepreneurial institution", implies a reconfiguration of technoscientific and sociopolitical orders. Importantly, this transition does not unfold in an isomorphic fashion – the University simply transforming into a "neoliberal university" (Slaughter and Rhoades, 2000) – but is shaped by the institution's attempt to transform into an innovation agent while living up to its civic responsibilities. This mode of "institutional eclecticism", however, calls into question existing lines of accountability and necessitates a critical rethinking of the role of the modern university in regional development.

Three key insights from the previous chapter will shape the forthcoming analysis. First, innovation spaces are inevitably *political* spaces. Technoscientific progress cannot be separated from political intent. Second, the "Gang of Four" (GoF) – understood as representatives of institutionalised scientific expertise – seek to maintain the appearance of apoliticism by mobilising three modes of apoliticisation: language, location and data science. Finally, global HEIs such as the University of Edinburgh are subject to internal and external transformations, which lead them to assuming an increasingly prominent role in the politics of urban-regional future-making.

This chapter is divided into four parts. I begin by reviewing existing literature in STS, the sociology of higher education and innovation studies, tackling the relationship between HEIs and urban-regional development. Afterwards, I look at how the re-emergence of the city region as a relevant spatial reference point shapes urban-regional governance restructurings. Having elaborated on the intimate relationship between scale and imagination, Section 4 will then critically study the tension-ridden relationship of three dominant imaginaries of the modern university: "ivory tower", "economic agent" and "civic institution".

2. From Humboldtian ideal to economic engine

This chapter contributes to existing academic literature in STS and the sociology of higher education, which examines the relationship between higher education and urban-regional development. Three themes in this body of literature build the foundation of the forthcoming argument: 1) the rise of HEIs as drivers of economic growth (Etzkowitz and Leydesdorff, 1995; Etzkowitz, 2003;

Arbo and Benneworth, 2007; Uyarra, 2010; Berman, 2012a; Audretsch, 2014); 2) the multi-scalar nature of universities (Power and Malmberg, 2008; Addie, Keil and Olds, 2015; Lee and Clarke, 2019); and 3) the notion of the "civic university" (Chatterton and Goddard, 2003; Goddard, 2009; Goddard and Vallance, 2013; Goddard *et al.*, 2014; Whyte, 2016).

As for the first strand, scholars in the sociology of higher education, such as Berman (2012a) and Audretsch (2014), emphasise how the role of Western HEIs evolved from being a resource to policy and economy towards being an engine of economic growth. Berman (2012a) argues that there are two main reasons why academic science moved towards the market. First, government policies encouraged the commercialisation of scientific practices and, second, policymakers started to identify scientific and technological innovation as key drivers of the economy. Similarly, existing innovation policy studies on the rise of the entrepreneurial university (Clark, 1998; Tornatzky, Waugaman and O'Gray, 2002; Thorp and Goldstein, 2013) and the "triple helix" model (Etzkowitz and Leydesdorff, 1995; Etzkowitz, 2003) highlight the roles of universities as key engines in regional and national development. No longer just acting as support structures for innovation by training people and providing knowledge to industry, the university is seen as part of a "triple helix" (university-industry-government) and as a "natural incubator" (Etzkowitz, 2003, p. 325) in knowledge-based economics.

While providing helpful insights regarding the synergistic interplay of government, industry and higher education, the contributions mentioned above conceptualise the process of innovation predominantly as devoid of political intent. Recent studies at the intersection of STS and innovation policy challenge the seemingly apolitical nature of innovation by emphasising that technoscientific and sociopolitical orders are always co-produced (Pfotenhauer and Jasanoff, 2017b; Pfotenhauer and Juhl, 2017; Pfotenhauer *et al.*, 2022). Pfotenhauer and Juhl (2017) stress that "innovations are at once technical and legal-political propositions that enter into dialogue with existing sociotechnical settlements first and foremost on normative grounds, not just as new artifacts to which legal and political frameworks merely need to adapt" (p. 82). In this chapter, I want to build on these insights and examine how innovation – inevitably charged with ideas about how the city region *ought* to develop – shapes imaginaries of the University of Edinburgh.

A second body of literature problematises the multi-scalar nature of HEIs. Goddard and Vallance (2013) point out that universities constitute "place-based institutions with connections into the different social and institutional spheres of [their] locality" and act as "generative node[s] in national and international flows of knowledge and people (especially highly-mobile students)" (p. 2). Unlike municipal authorities, which are bound to specific geographies, global HEIs such as Edinburgh University operate across local, regional, national and global scales. It follows that universities' spatial outlooks might not necessarily be local but rather shaped by national and international innovation landscapes. In other words, while universities are generally considered to contribute to innovation, it is far less clear how they impact *urban-regional* innovation (Power and Malmberg, 2008; Addie, Keil and Olds, 2015; Lee and Clarke, 2019). Therefore, it is important to scrutinise existing models of innovation, which tend to assume that institutional hybridisation inevitably leads to innovation. Instead, this chapter asks: 1) what role do universities actually play in urban-regional innovation systems? and 2) how does their multi-scalar nature affect various interpretations of the modern university?

Finally, one strand of literature attempts to complicate and enrich existing conceptualisations of the role of HEIs in urban-regional development by engaging with the notion of the "civic university". Seeking to break down barriers between universities and cities, the concept of the "civic university" indicates a shift from the "triple helix" to the "quadruple helix" (government, business, higher education, *civil society*). Goddard and Vallance (2011) critique existing literature in regional development for narrowing the function of universities down to "source[s] of knowledge generation and dissemination within a local economic innovation system" (p. 433). Instead, they suggest that the university – understood as a *civic* institution – should place its research and education in service of the community by engaging as a whole with its surroundings, partnering up with other local universities and using its location to form its identity (Goddard, Vallance and Kempton, 2012).

It is important to note that in the UK context, the notion of the civic university by no means constitutes a recent phenomenon. For instance, during the Victorian era, a first wave of so-called "Redbrick" universities emerged as centres of knowledge and research (Whyte, 2016). Originally comprising the universities of Leeds, Bristol, Liverpool, Birmingham, Manchester and Sheffield, Britain's civic universities were modelled after the Humboldt University of Berlin and constituted a response to the societal pressures of the Industrial Revolution. In what follows, I want to ask: Which actors mobilise the imaginary of the *civic* university when describing the role of the University of Edinburgh in the city region? What notions of citizenship are invoked? Who is imagined to be part of the citizenry and who is not?

In summary, his chapter builds on existing academic debates at the intersections of STS, the sociology of higher education and innovation studies, which examine the role of universities in urban-regional innovation economies. More specifically, it seeks to explore which competing – if any – imaginaries of the University of Edinburgh are mobilised and negotiated by whom along two suggested axes: "local-global" and "open-closed".

3. Scale and imagination

> "Place. That's key. So if you look at all the evidence of the last couple years, the thing that tends to make a difference is place. People care about place. And if you look at all the ones [universities] that are successful around the world, they have tended to focus around an area." (Interview 34, 2020)

This statement was made by a senior Edinburgh University official responsible for the institution's strategic partnerships. Reflecting on the role of the university in data-driven economic growth, they evangelised about the need for "place-based innovation". The latter – one of the latest technoscientific "buzzwords" (Bensaude Vincent, 2014) – underscores the importance of mobilising physical proximity and local knowledges when driving innovation (McCann and Soete, 2020; Department for Business Energy and Industrial Strategy, 2021; UK Government, 2021). It is much less clear what the notion of "place-based innovation" actually entails: in relation to what *place* is the University imagined? Before elaborating on different imaginaries of the 21st century university, it seems crucial to focus on the context in which different interpretations of city region emerge. In this section, I contend that the *city region* has (re)emerged as the dominant socio-spatial reference point for reimagining the modern university. Putting forward this line of argument, I will first explore the relationship between state re-scaling processes and urban-regional governance restructuring and, second, critically engage with the idea of the "knowledge economy".

3.1. Rediscovering the city region, reimagining the innovation state

As highlighted in previous chapters, the emergence and implementation of Edinburgh's innovation apparatus must be understood against the backdrop of the continual restructuring of regional development policy in the UK (Brenner, 1998; O'Brien and Pike, 2015; Martin *et al.*, 2019). Martin *et al.* (2019) point out that "the territorial focus of subnational economic development policy has changed frequently from regionalism to localism, back to regionalism, then localism and most recently to city-regionalism" (p. 52). Keeping the frequent spatial shifting in mind, one cannot help but notice that scale and imagination are intimately tied together. Put differently, the "politics of scale" shapes the "politics of the imagination".

Importantly, by "politics of scale", I am *not* referring to the "scalability zeitgeist" or the "politics of scaling" (Pfotenhauer *et al.*, 2022). Rather than linking scale to the notion of "up-scaling" – which dominates discourses, for instance, around the emergence of platform technologies – "politics of scale" implies that depending on the chosen geographical scale, different imaginaries can be invoked. Scale functions as a framing device, as a wallpaper that shapes actors' conceptualisations of the self. In line with the co-productionist research agenda on the innovation state (Pfotenhauer and Juhl, 2017), I contend that the UK and Scottish governments do not simply act as facilitators of innovation-based economic growth but actively structure actors' self-imaginations. In recognising and elevating the city region as *the* relevant socio-economic space of innovation, the *political* state sets out and mediates the boundaries within which innovation agents such as Edinburgh University can imagine themselves.

This observation must be further complicated by recognising the inherently co-productionist nature of state re-scaling processes. It is not just the state that produces new spaces of innovation. Innovation spaces also help reimagine the modern innovation state. Studying the relationship between state re-scaling and urban-regional governance restructuring, Brenner (2019) points out that urban space and state space are "intricately entangled, mutually co-constituting and conflictually coevolving formations of scale-differentiated sociospatial relations under modern capitalism" (p. 10). It follows that defining the Edinburgh city region as the "DCoE" represents more than the result of a unidirectional imagination exercise by the state. As much as the UK and Scottish governments – via policies such as city region and growth deals – create relevant units of socioeconomic spatiality, it is also proclaimed innovation spaces which help imagine Scotland as a modern nation or what Moisio and Rossi (2020) call a "start-up state". The Scottish government's most recent £45m plan to become a "tech nation[1]" by "transform[ing] Scotland into one of Europe's leading economies to start or grow a technology business" (Scottish Government, 2022b, p. 1) is testament to this co-constitutive process.

Moving forward there are at least two key insights emerging from the previous elaborations. First, self-imagination processes never take place in a vacuum. Not only does context matter but context itself is subject to frequent change. Studying the Edinburgh city region, it becomes clear that state scaling and re-scaling processes shape urban-regional restructuring arrangements. In other words, different sociospatial configurations implicate different landscapes of imagination. Rather than just constituting a passive bystander, the state actively mediates and engages in the "politics of scale". Second, state space and city-regional space are co-constitutive (Brenner, 2004, 2019). Brenner (2019) aptly remarks that re-scaling processes, taking place at the city-regional and state level, must be analysed "as mutually intertwined processes of sociospatial restructuring" (p. 117). (Re)imagining the Edinburgh city region into the DCoE simultaneously implies (re)imagining Scotland as a nation built around data-driven innovation and entrepreneurship.

To summarise, it is crucial to take into account *scale* and *scaling* as helpful epistemologies when analysing how competing imaginaries of the 21st century university emerge and stabilise. Importantly, adopting a relational and multiscalar understanding of politico-economic restructuring processes does not deny the existence of "processes of sociospatial agglomeration" and "highly consequential zones of concentration and intensification" (Brenner, 2019, p. 112). Rather, it implies resisting the temptation to conceptualise the city region as a self-enclosed and bounded area of socioeconomic activity. Instead, it must be grasped as "actively constituted and continually transformed through the constitutively multiscalar processes, crisis tendencies, strategies, struggles, and conflicts associated with the problem of territorial organization under capitalism" (Brenner, 2019, pp. 169–170).

3.2. The "knowledge economy"

Before engaging with self-imaginations of the modern university in more detail, it makes sense to further untangle the relationship between regionalisation, innovation and higher education in the UK. Starting with New Labour

---

1 More information can be found here: https://www.gov.scot/news/building-a-tech-nation/.

in the mid-1990s, the *region* has (re)emerged as a relevant spatial reference point and a key economic unit. Uyarra (2010) observes that "the interest in knowledge-related regional university-industry networks coincides with a policy and governance trend towards regionalization of innovation and knowledge exploitation activities" (p. 1237). The increasing importance of place in science and innovation policy was not only reflected in the emergence of Regional development agencies (RDAs)[2] in Scotland (1991) and England (1998), or city region and growth deals across the UK from 2010 onwards. More generally, it was echoed in a zeitgeist that preached the gospel of the "knowledge economy" (Florida, 2012; HM Treasury & Department for Business Innovation and Skills, 2014; Flanagan and Wilsdon, 2017).

The concept of the "knowledge economy" attempts to capture the transition from land, labour and capital to an economy dominated by knowledge-intensive activity. One notion, strongly pushed by proponents of the knowledge economy, represents the idea of the "ideopolis". The "ideopolis" or "knowledge city region" embodies the vision of a place characterised by a highly educated labour force, the strong presence of universities, as well as efficient communications and transport infrastructures (Jones *et al.*, 2006). One highly influential policy, which can be read as a direct response to the "new demands of the knowledge economy" constitutes the launch of the "Northern Powerhouse" (HM Treasury, 2016). Introduced during a speech at the Museum of Science and Industry in Manchester in 2014, Then-Chancellor of the Exchequer, George Osborne pointed out: "We want to see science here turned into products here – and into jobs and growth here" (Osborne, 2014, p. 1).

Invoking the role of science and technology in "placemaking", Osborne emphasised the importance of city regions as key vehicles of economic growth. Subsequently, the UK government – in both the Treasury's summer budget 2015 (HM Treasury, 2015) and Jo Johnson's "One Nation Science" speech (Johnson, 2015) – further specified its vision for science in regional economic development by launching the Science and Innovation Audits, as mentioned in Chapter 5. From the beginning, policymakers assigned great importance to HEIs as central innovation agents. Given their local rootedness, universities

2 As part of austerity measures, RDAs were abolished in England in 2010 and partly replaced by local enterprise partnerships. In Scotland, there still exist two economic and community development agencies: Scottish Enterprise and Highlands and Islands Enterprise.

were expected to act as "anchor institutions", which move beyond their traditional missions of teaching and research (Charles, 2003, 2006; OECD, 2007; Huggins and Johnston, 2009; Flanagan and Wilsdon, 2017). A 2007 OECD report on "Higher Education and Regions" stresses that

> "To be able to play their regional role, HEIs must do more than simply educate and research – they must engage with others in their regions, provide opportunities for lifelong learning and contribute to the development of knowledge-intensive jobs which will enable graduates to find local employment and remain in their communities" (OECD, 2007, p. 11)

Building on the notion of HEIs as urban-regional "anchor institutions", Perry (2008) identifies three dimensions of Knowledge-based urban development (KBUD). The *process*-driven view of knowledge-based development illuminates new processes of knowledge production and conceptualises universities as critical actors who provide strategic intelligence. This perspective asks "fundamental questions relating to how knowledge is produced, for what reasons, by whom, for whom, and how it is subsequently judged" (Perry, 2008, p. 24). A second dimension – the *product*-driven view – revolves around the exploitation of particular knowledge products and assigns a passive, facilitative role to HEIs. Here, the production of knowledge is separated from its application.

Finally, the *acquisition*-driven view of KBUD builds on the notion of the "ideopolis" and conceptualises universities as one among many partners in urban-regional development. Studying the emergence and implementation of the "Manchester: Knowledge Capital" (M:KC) initiative, Perry (2008) argues that this third view constitutes "a lowest common denominator approach" (p. 34). She points out that the M:KC initiative was characterised by "a strong focus on physical development, estate management, and developing world-class universities and places through attracting knowledge workers and developing a cultural buzz" (Perry, 2008, p. 33). Put differently, instead of engaging with the complex processes of knowledge production and dissemination, policymakers fostered partnerships and consensus around physical developments.

This realisation is echoed by Lawton-Smith (2007) who cautions against ascribing too much importance to universities' contributions in regional innovation economies. Critically engaging with five propositions as to why universities constitute crucial stakeholders in innovation-driven economic development, she argues that universities are an "overstated ingredient in determining the location of R&D and high-technology industry" (Lawton-Smith, 2007,

p. 111). Instead, she contends that it is political agendas and particularly government spending that shape regional development. Engaging with this body of literature indicates that the relationship between regionalisation, innovation and universities is much more tension-ridden than notions such as the "knowledge economy" or "ideopolis" suggest.

To sum up this section, the (re)emergence of the city region as a relevant spatial reference point is the result of both state re-scaling processes and emergent ideas of an economy built around knowledge as a key driver of growth. Importantly, one cannot think one without the other. Ideas of technoscientific progress are co-produced with sociopolitical orders. Critiquing realist ideologies which consistently separate facts from values and reason from politics, the idiom of co-production emphasises that the ways in which we choose to produce and govern technoscientific knowledge are inseparable from the ways in which we choose to live in society (Jasanoff, 2004). Accepting this line of argumentation, I am now in a position to examine competing imaginaries of the 21st century university. Central to this endeavour is the following question: What does it mean to imagine the university in relation to the *city region*?

4. Competing imaginaries of the 21st century university

Having elaborated on the co-constitutive, multiscalar and dynamic context in which relevant spaces of innovation emerge, this section will focus on the university as a key agent in the social organisation of knowledge production, dissemination and stabilisation. Doing so, I ask: what is at stake when imagining the university against the backdrop of state re-scaling processes on the one hand and increasing integration into global innovation landscapes on the other? Which imaginaries of the modern university are at play when reconceptualising the Edinburgh city region as the DCoE? How do these multiple interpretations work out in practice?

In what follows, I argue that while there is a clear shift from imagining the university as an "ivory tower" to grasping it as a globally networked, entrepreneurial institution, the University of Edinburgh – given its status as Scotland's premier university – attempts to accommodate a variety of roles. More specifically, there are three dominant imaginaries: the "ivory tower", the "civic institution" and the "economic agent". Imagining the university as a networked, entrepreneurial organisation does not imply erasing any of the other two prevalent imaginaries. Instead, the university engages in a mode of "institutional eclecticism" which reveals inherent frictions as to what the university *should* and *should not* do.

4.1. The "ivory tower"

One imaginary of the modern university persistently invoked when asking respondents about the role of HEIs in urban-regional development represents the notion of the "ivory tower". The latter – also referred to as the Humboldtian ideal – "describes the university as an autonomous body of self-governing professionals, accountable to and monitored by itself" (Baert and Shipman, 2005, p. 159). Supposedly separated from politics, economy and civil society, the pursuit of knowledge is understood as valuable in and of itself, since it directly feeds into societal ideals such as *Bildung* and self-edification, referring to self-cultivation and self-improvement (Audretsch, 2014).

Importantly, while we can acknowledge that these intellectual activities might lead to practical spill-overs, it is neither explicitly expected nor intended. This particular understanding of the modern university is most clearly emphasised by actors *outside* the university which reinforces the insular nature of the institution. The following quote, by one senior official working at one of Scotland's regional development agencies, serves as a good example.

> "The big universities like the University of Edinburgh, and it's changed slightly, but we would always regard those as pretty irrelevant to the economy. They have produced a few start-ups, they have collaborations with global giants but it doesn't affect the day-to-day industry in Scotland. We see them as sort of astride the world but 'ignoring their own backyard', is the phrase I use. They love to be global players doing things all over the place, but they couldn't name six companies around the corner from them." (Interview 23, 2020)

Questioning Edinburgh University's assigned role in leading urban-regional innovation, this individual goes on to argue that it should be further education colleges, such as Edinburgh College, or industry-oriented universities, such as the University of Strathclyde, who should be driving innovation-driven growth. This sentiment is shared by a chief executive representing local business interests. When questioned about the involvement of the business community in the university's regional development plans, this official responds:

> "And I am not sure, if I am honest, there is a level of arrogance that they can do everything better and that they can do it themselves. And I am sure the penny will drop that we are a fantastic resource that could cover a lot of ground much faster." (Interview 17, 2020)

A high-ranking third sector representative mirrors this impression and further problematises the exclusionary institutional culture of HEIs:

> "I'm not sure how accessible any of the universities feel to most people, especially people who haven't had an education, an university education. I think universities, like all educational institutions, are quite intimidating to people. If you haven't had experience of that and a positive experience of your education journey. So universities or any further education institution can become quite a rarefied atmosphere and it's quite easy to speak in a bubble." (Interview 37, 2020)

Recalling the findings of an Audit Scotland (2020) report, which concludes that local communities have had very little involvement in the CRD, Edinburgh University has yet to prove how it intends to serve the wider community. Reflecting on who mobilises the – predominantly negatively connotated – imaginary of the "ivory tower", it becomes clear that it is above all those actors who have been delegated neither financial resources nor ideational responsibility: third sector organisations and regional development agencies.

4.2. The economic agent

A second "collectively held, institutionally stabilized and publicly performed" imaginary (Jasanoff and Kim, 2015, p. 4) captures notions of the university as a highly networked, economic agent. Invoking what Etzkowitz and Leydesdorff (1995) call the "triple helix model of innovation". The University of Edinburgh is perceived as opening up to industry and government, and imagined as a key driver of future city-regional growth. Berman (2012a) observes that "the shift from a 'science-as-resource' to 'science-as-engine' model […] gave universities a new mission: to facilitate economic growth by making sure their research reached the marketplace" (p. 3). This "shift" is recognised by the majority of my interviewees. This local business representative, for instance, points out that

> "now that they [Edinburgh University] are investing in spin out businesses, they got the Edinburgh Innovations[3] which is supporting businesses spin out and then develop into commercial entities. And they are actually investing and supporting businesses through that. You know I do see the university as really opening up as a strong and key contributor to the city." (Interview 17, 2020)

The university's ambitions to transform into an entrepreneurial institution are intimately tied to a 2020 review on Scotland's technology sector, conducted by Mark Logan, former Skyscanner executive and Professor of Computing Science at the University of Glasgow. The Logan Review (2020) identifies a discrepancy between knowledge generation and knowledge application and assigns special importance to universities to close that gap. More specifically, it stresses that

> "The best universities view industrial collaboration in general, and start-up facilitation, in particular, as equally important to their other missions of teaching and research. They understand that it is difficult to be genuinely world-class in teaching and research without being world-class in industrial liaison and entrepreneurial support. With any of these missing the others are diminished." (Logan Review, 2020, p. 29)

Unsurprisingly, key decision-makers and specifically members of the GoF, not only share this entrepreneurial vision of the university but view it as a lifeline to stay globally competitive. One high-ranking official emphasises

> "I think we've seen the university start to open up. [...] I think there's a recognition that we're not as good as we think we are. I think this is starting to have a recognition that there's a high degree of complacency and that we're over relying on our reputation. [...] We need a good kick up the arse if I'm honest, as a university if we want to remain global top 20. And I think, you know, data-driven innovation, the work we're doing in Fife, etc., is demonstrating just the first indications of the impact that we can have on the Scottish economy and to some social impact." (Interview 21, 2020)

3 More information can be found here: https://edinburgh-innovations.ed.ac.uk

One key architect behind the DDI Programme echoes these remarks and highlights the importance of best-practice models of innovation such as Silicon Valley:

> "The argument is relatively simple in the sense of saying, and it is around the strapline of Data Capital of Europe. The ability to attract and retain interest from outside the region would have spill-over effects within the region. So you kind of again, you go back to places like Stanford and Silicon Valley and that type of argument. So we are the initial driver to attract people. We are the driver to attract students to start-ups. We will have this supposed ecosystem which will include investors and so on." (Interview 44, 2020)

Edinburgh University – portrayed as a globally networked, entrepreneurial institution – is imagined as a regional anchor institution that – by attracting talent, funding and companies – secures future economic prosperity. The university's view across the pond is by no means surprising. Examining "travelling imaginaries of innovation", Pfotenhauer and Jasanoff (2017b) identify three best practice models: the MIT model, the Silicon Valley model and Responsible Innovation. Studying the emergence of the CRD, it becomes clear that the Silicon Valley model serves as the dominant framework for envisaging the modern university. This particular notion of innovation locates successful innovation at the regional level and "include[s] a mix of excellent academic institutions, technology industries, venture capital, close networks of highly skilled individuals, and a risk-taking mentality that together drive an abundance of start-ups and multinational companies" (Pfotenhauer and Jasanoff, 2017b, p. 420).

4.3. The civic institution

The concept of the "civic university" constitutes a third imaginary of the modern university that actors in the Edinburgh city region consistently mobilise. As mentioned above, imagining the university as a civic institution promises to move beyond simplistic innovation models such as the "triple helix", and critically engage with what it means for HEIs to transition from being *in* the city to being part *of* the city. Linking the university's efforts to fulfil its civic responsibilities to the vision of transforming the city region into the DCoE, McAra (2019) urges that "if we are to realise the promise of data-driven innovation for society and mitigate against its darker potential, this will require more effective collaboration and critical friendship between universities, governments, industry and the communities within which they are situated and whom they serve" (p. 46).

In line with the previous two imaginaries, this imaginary of the "civic university" is invoked and shaped dissimilarly by different actors. The following two quotes illustrate two contrasting perceptions of "civic" when reflecting on the role of Edinburgh University in implementing the CRD. Once again, it becomes obvious that there is not only a discrepancy between self-perception and external perception but more generally between global HEIs and local community representatives. While the higher education official classifies the institution's involvement as "giving back *to* the city", the NGO director assesses the university's outreach activities as continued "taking *from* the city":

> "So I think we were too introverted and we didn't think about what was our civic role. And these are words that [name], [name] and I use and others. We got a duty to the city. We were one of the first civic universities and we'd forgotten that. Given the scale of the university in the international region and our credibility by being a world 20 top university, it gives us an authority and an evidence base to say 'no, this stuff works'." (Interview 34, 2020)

> "I don't see a huge amount of reaching out to other sectors, apart from sectors which they would also feel are influential. I think that is a continuous frustration for me is that they're reaching out to other influencers rather than to those who haven't got the same resource or the same capacity to influence. And actually, that means you miss out on a lot of things and a lot of possibilities, opportunities." (Interview 30, 2020)

One cannot help but notice the interpretative flexibility inherent to the notion of the "civic". More generally, these two statements demonstrate that notions of citizenship are invoked dissimilarly in different contexts. It follows that it is crucial to critically engage with who is imagined to be part of the urban citizenry and who is not.

4.4. Clashing ideals

Having introduced three dominant imaginaries that shape the perception of the University of Edinburgh in the context of the CRD, I now want to engage more deeply with what is at stake in these identity-making practices. What are the implications of imagining the university as a globally networked, entrepreneurial institution? How does this vision align and/or compete with notions such as the "civic institution" or "ivory tower"?

From the outset, it becomes clear that reimagining the 21st century (urban-regional) university is by no means an easy task. On the contrary, the relationship between regionalisation, innovation and universities is much more tension-ridden than notions such as "knowledge economy" or "ideopolis" suggest. University senior leadership's recent ambitions to drive data-driven innovation bring to light inherent tensions between self-imaginations as a networked, entrepreneurial institution and more community-driven and traditional imaginaries of the university.

Reflecting on the UK and Scottish governments' investment into the CRD, this high-ranking official from one of Scotland's economic development agencies pointedly remarks

> "The government is taking a bit of a gamble here, giving so much money to these [universities], essentially building buildings and those kinds of things with all the capitals and in the city region deal. So the constant question is, okay, you build that building. How does that change industry? I don't think there are very many convincing answers. It's all a sort of a 'wait and see mode'. So it may be a good way of helping the digital economy or it may be rubbish. We don't know yet." (Interview 23, 2020)

This statement reveals at least two relevant insights. First, it demonstrates a healthy scepticism from the "old guard" of regional development – regional development agencies such as Scottish Enterprise – towards emergent innovation agents such as Edinburgh University. Second, this line of critique is also recognised in a body of literature which critically examines the multi-scalar nature of HEIs and particularly universities' contributions to *regional* innovation (Lawton-Smith, 2007; Perry, 2008; Power and Malmberg, 2008). Acknowledging that universities can bring material and immaterial advantages, such as spin-off companies or relevant labour skills, to their host regions, Power and Malmberg (2008) underscore that one should not expect "the world-class research carried out in universities to result in innovations that get exploited as economic activities" (p. 243).

This finding is in line with Perry's (2008) study on the M:KC initiative, mentioned in Section 3.2, in which she argues that universities are not necessarily driving new processes of knowledge production but are rather fostering partnerships around physical developments. Crucially, the Manchester case study shows strong similarities with the CRD: the DCoE vision is literally *built* around the six innovation hubs. These remarks beg the following questions: Will the university get into a position where it can provide economic value beyond "accidental contributions" such as student accommodation and consumption in the city region? Is it enough to rely on spill-over effects given significant government investment into the university?

A second fundamental tension that is persistently pointed at by many of my interviewees refers to the interplay of the "entrepreneurial" and the "civic". Elaborating on the emergence of the CRD, one NGO director sharply emphasises that the CRD felt "like a spaceship that landed" (Interview 40, 2020). Questioned about potential issues around democratic legitimacy, they go on to say

> "All of a sudden, there is a city region deal. There's money. And you know what, it's going to x, y and z. That's great but who decided that? Who decided what the priorities are with all this money? Who decided that? Because I didn't. I don't know anybody else that did. It just landed on us." (Interview 40, 2020)

Interestingly, this accusation is not even disputed by key decision-makers. Both higher education officials and senior policy-makers acknowledge that the CRD "is not a piece of democracy" (Interview 46, 2020), and that "the way in which these deals are struck together don't allow for much community stakeholder engagement input" (Interview 8, 2020). Playing with the spaceship metaphor, a chief executive officer of one of the six councils goes as far as saying that "it depends whether you were in the spaceship when it landed or whether you were looking at it when it landed. So I was inside looking out" (Interview 48, 2020). The following statement by the same individual captures a dominant sentiment among key decision-makers when reflecting on their involvement in the CRD:

> "I don't think there was much in the Edinburgh and South-East city deal that was genuinely set out to be community driven and community focused. It was set out to be community beneficial, beneficial to the community. It was set out to tackle deprivation, vulnerability, social deprivation. [...] So do you see what I'm doing? I am saying there is a connection between the knowledge of those who are doing the development and their understanding and the setting in which this work was going to be done." (Interview 48, 2020)

Is it enough to excuse the lack of community engagement by claiming that councils know their residents well enough to make decisions without them? This assumption seems highly problematic given the scale – £1.3bn – and the transformational potential and promises attached to the CRD. Not only does it reveal hubris and a lack of democratic sensibility but, most strikingly, it showcases an understanding of innovation that is devoid of politics and, instead, rooted in techno solutionism. That is, an almost blind belief in the power of digital technologies to solve highly complex social problems. Asked about how the "woman on the street" can relate to data-driven innovation, this council leader points out

> "I think a better question is 'Does it matter? Does it matter that most people in the city don't know what DDI is? Or is that important or not important?' I don't know if it's important or not. I think the important bit is, will it drive solutions that people will understand?" (Interview 1, 2019)

One criticism that might be levelled at this statement is that it denies citizens any interest in or knowledge of digital technologies and innovation. It reinforces the techno-solutionist standpoint by adopting an instrumental understanding of technology. In other words, data-driven innovation simply constitutes a means to an end: "if DDI happens to be the latest driver of economic growth so be it". Most importantly, this individual separates technology and innovation from society and politics. They do not acknowledge that technology and innovation are inevitably co-produced with norms and values about how society *ought* to develop. Denying the sociopolitical embeddedness of technology-based innovation directly feeds into the apoliticisation of technoscientific development mentioned in the previous chapter.

Elaborating on the tension-ridden relationship between entrepreneurial, civic and isolationist imaginaries of the university leads to a sobering interim conclusion: it is far from clear whether the University of Edinburgh – via its engagement in data-driven innovation – can positively impact either the regional economy or local communities. Instead, Edinburgh's ambitions around data-driven innovation – with a strong "entrepreneurial university" at the heart of it – reflect a techno-solutionist zeitgeist that is shaped by its top-down and exclusionary nature.

4.5. Institutional eclecticism

> Was it intended to be a college devoted to the liberal arts or a Protestant seminary, designed to produce much-needed recruit for the parish ministry? Was it to be a new-style college or old-style university? Was it to be a college for the town, or an institution meant to attract students from afar? The irony is that the scheme was, at one time or another in its pre-history, all of these things. (Lynch, 2003, p. 6)

Documenting the early history of the University of Edinburgh – or, as it was known in the 16th century, "Tounis College" – Michael Lynch shows that since its foundation in 1583, the University's purpose was never clear. Further complicating the analysis of identity-making practices in the previous sections, I now want to zoom in on how the three competing imaginaries play out *within* the university. Following Barrioluengo *et al.* (2019), who argue against the claim that macro factors inevitably lead to institutional isomorphism, I contend that the University of Edinburgh attempts to accommodate a range of different imaginaries in order to legitimate its status as Scotland's premier university. This mode of "institutional eclecticism" materialises itself most prominently amongst the six DDI innovation hubs, which embody different imaginaries of the modern university (Figure 29).

Figure 29 locates the six innovation hubs in relation to the three dominant imaginaries of Edinburgh University along two axes: "closed-open" and "local-global". Before engaging more closely with the individual innovation hubs, it is important to clarify some of the key assumptions underlying this figure. First, global HEIs such as Edinburgh University are inherently multi-scalar institutions. Unlike city councils or local NGOs, they are not geographically bounded but operate across local, regional, national and global scales. Taking this observation as an analytical starting point, it is, nevertheless, possible to locate the university along socio-spatial ideal types (ivory tower; civic institution; economic agent) portraying what Brenner calls "highly consequential zones of concentration and intensification" (Brenner, 2019, p. 112).

Figure 29: Competing imaginaries of the 21st century university | Source: Own illustration

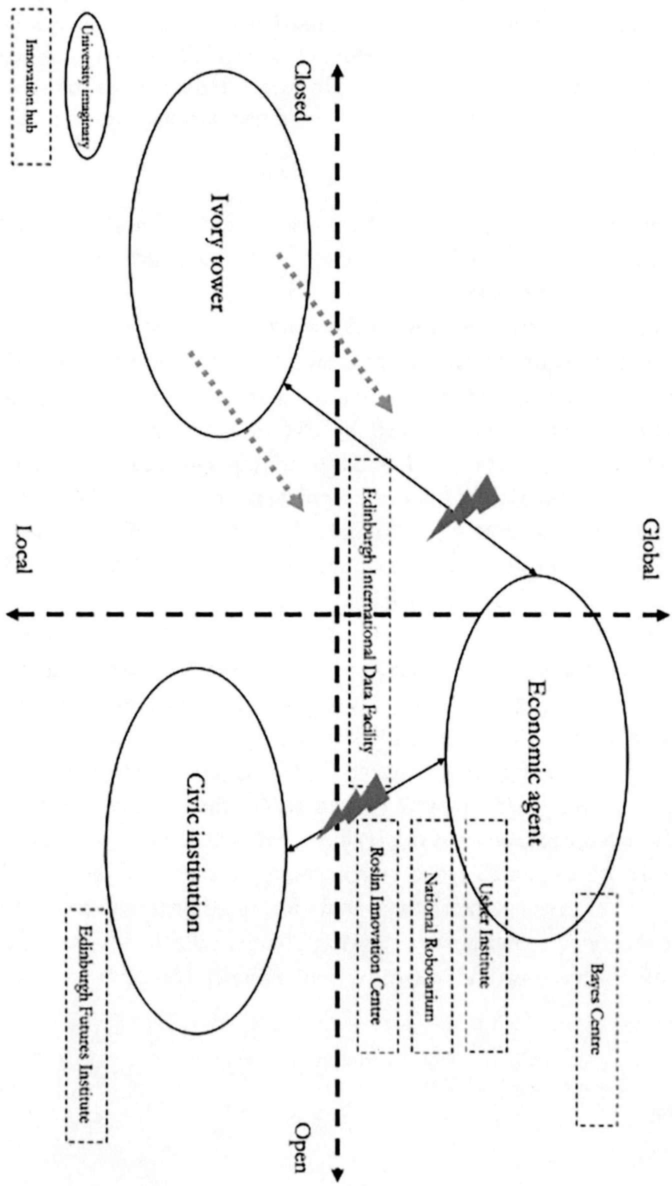

Second, based on these considerations, the "local-global" axis illustrates the degree of geographical outreach while the "open-closed" axis represents the degree of porosity toward stakeholders outside the university walls. The example of the Edinburgh Futures Institute (EFI) illustrates this classification. While embedded in international research networks (the multi-scalar nature of HEIs), the EFI intends to embody the imaginary of the civic, porous university, seeking to listen to and invite local communities in the production of scientific knowledge. Finally, Figure 29 inevitably constitutes a synoptic representation of the knowledge games taking place in the context of the CRD. It is important to mention that the three imaginaries are neither mutually exclusive nor static but mobilised dissimilarly by different actors in changing contexts.

Having established some basic premises, I now want to elaborate on how the six DDI innovation hubs move within this "playground of imagination". From the outset, it becomes clear that none of the innovation hubs neatly fits just one of the three university imaginaries. Instead, each draws on and embodies civic, entrepreneurial and more isolationist understandings to varying degrees. While in line with the general trend of transitioning from "closed-local" to "open-global" there are noticeable differences.

The two innovation hubs that come closest to resembling a socio-spatial ideal type are the EFI and the Bayes Centre. While the former aspires to live up to the ideal of civic institution, the latter – the university's innovation hub for Data Science and Artificial Intelligence – is emblematic of what Zukin (2020d) calls the "urbanisation of Silicon Valley" (p. 3). Bayes and EFI epitomise very different understandings of how the University of Edinburgh should drive data-driven innovation. Backed by the College of Arts, Humanities and Social Sciences, the EFI attempts to adopt a more critical attitude towards what evidence-based policy looks like. Embracing the Geddian notion of "civics"[4]

4 The notion of "civics" or the "theory of civics" represents Patrick Geddes' attempt to bring sociological analysis into conversation with town planning. Defining civics as "applied sociology", Geddes (1904) specifies: "Viewed as Science, Civics is that branch of Sociology which deals with Cities, their origin and distribution; their development and structure; their functioning, internal and external, material and psychological; their evolution, individual and associated. Viewed again from the practical side, that of applied science, Civics must develop through experimental endeavour into the more and more effective Art of enhancing the life of the city and of advancing its evolution. With the first of these lines of study, the concretely scientific, our philosophical outlook will not fail to widen; with the second, the practical, our ethical insight will not fail to deepen also." (p. 111)

(Geddes, 1904), it strives to be a place to study and learn about the city. Rather than deciding what urban futures are desirable it seeks to adopt an ontology of the urban that allows to investigate the virtue of being "civic". As a result, it seeks to promote epistemologies from the social sciences that allow for situational, experiential and local knowledges.

By contrast, the Bayes Centre values different forms of knowledge-making. Backed by the University's College of Science and Engineering, it embodies and performs a form of higher education-driven urban development that is much closer to the techno-solutionist zeitgeist mentioned in Section 4.4. Inspired by global reference points such as Stanford and Silicon Valley, the Bayes Centre mobilises data science as a *method* which allows to *describe* hidden layers of urban-regional life. Instead of viewing data science as an "organising idea" (McQuillan, 2018) that *shapes* sociotechnical realities, Bayes adopts an instrumental approach and envisions an urban-regional society comprised of entrepreneurs and service-users. The city region is imagined as a set of data points which allows the university to *solve* complex social problems. To put it in a nutshell, whereas the EFI aspires to explore multi-faceted and local ontologies of the urban, the Bayes Centre builds on fixed ontologies and applies epistemologies, which are rooted in global tech and innovation discourses.

The remaining four innovation hubs – Roslin, Usher, EIDF and Robotarium – embody more hybrid imaginaries of the 21st century university. While confirming the general shift from "ivory tower" to "economic agent", there are differences pertaining to desired audiences. In other words, it is one thing to "open up" – the more interesting question is: open to whom? Here, the four DDI hubs put forward disparate conceptualisations of relevant publics. The Roslin Innovation Centre and the National Robotarium embody traditional higher education third mission activity by strongly emphasising industry engagement. Seeking to commercialise their research, both hubs promote themselves as "business gateways". The EIDF – providing the backbone of the wider DDI initiative – addresses both academic and industry audiences. As much as they seek to encourage business collaboration, one key motivation is to attract scholars from all over the world to conduct data-driven research (Interview 19, 2020; Interview 29, 2020). Finally, the Usher Institute represents what could be called the imaginary of the "public service university". Given its focus on health and social care, it adopts a wider understanding of the public and occupies a prominent role in communicating science to non-expert audiences. One example of the latter constitutes "UNCOVER", the "Usher Network for Covid-19 Evidence Reviews"[5]. Reviewing and summarising relevant evidence in the context of the Covid-19 pandemic, this research network seeks to inform and guide the decisions of policy-makers.

To sum up, the university is changing. The reimagination of Edinburgh University occurs both *inside* and *outside* the institution. While roughly unfolding in the bounds of three dominant imaginaries – ivory tower, civic institution, economic agent – this process of identity and knowledge-making is messy and subject to endless friction. This statement by a high-ranking "agent of change" aptly captures the tension-ridden nature of this form of "institutional eclecticism":

> "You get many academics who say, I don't want to do high impact industry, public sector related research. It's not why I joined the university. I would have been a millionaire if I joined [the] industry habitat. I didn't want to. I want to just study my chosen subject. And I have my little office and I write research grants once every two years or three years. And I really focus on my chosen subject. And I don't really want to talk to anybody else apart from when I come to publish, and present my work. And that insular approach has got to go in my view. What we need is multi-disciplinary, cross college, cross schools programmes at scale in partnership with public sector and economic and social output. If I was [the] Scottish government and UK government. I would absolutely put that front and centre of every university in the UK agenda." (Interview 21, 2020)

Crucially, "institutional eclecticism" – the accommodation and interplay of various imaginaries – is shaped by and shaping sociopolitical orders. In other words, given its status as Scotland's premier university, located in the capital city, Edinburgh University is inevitably torn between transitioning into an entrepreneurial, networked and global institution, on the one hand, and remaining locally embedded as a citizen of the Edinburgh city region, on the other hand. This interplay of forces (local vs. global, open vs. closed) – under the guise of innovation – necessarily calls existing forms of organising knowledge into question. Building on this insight, I will, in the final section of this chapter, critically look at shifting lines of accountability resulting from the reimagination of the modern university.

---

5 For more information, see: https://www.ed.ac.uk/usher/uncover

4.6. Satellite or planet?

The three dominant imaginaries of the modern university – ivory tower, economic agent and civic institution – and their manifestations on the ground (e.g., innovation hubs) are representative of specific ways of producing and valuing knowledge. In other words, they encode different configurations of the knowledge-power dyad. Imagining Edinburgh University as a civic institution implies a divergent understanding of "truth-making" rather than imagining the university as a globally networked, entrepreneurial organisation. It does matter whether truth-making infrastructures such as Edinburgh University embrace *listening* and *observing* as opposed to *solving* and *deciding*, *bottom-up* as opposed to *top-down*, and *co-productionist* as opposed to *instrumental* epistemologies.

Based on the assumption that technoscientific and sociopolitical orders are co-produced (Jasanoff, 2004), the shift from isolationist imaginaries such as the "ivory tower" to globally networked imaginaries such as the "entrepreneurial university" can be read as signalling a transformation to a new social organisation of knowledge. The latter identifies the city region as a relevant unit of scientific activity, promotes innovation as a desirable mission and assigns projective agency to the university as a key driver of change. Building on Section 3, in which I elaborated on the co-constitutive relationship between the "politics of scale" and the "politics of imagination", it is crucial to problematise what happens when the University of Edinburgh is elevated from accountability to itself (ivory tower), to accountability to the city region (civic institution), to accountability to globalist reference points of innovation (entrepreneurial university).

The emergence of the CRD and its focus on higher education-driven growth reflects attempts to redefine the relationship between knowledge, power and society. The University of Edinburgh is moving from being an outsider to local and national politics to sitting at the heart of change. The recently passed "Regional Prosperity Framework", which constitutes a follow-up agreement directly building on the CRD, is testament to its enormous influence. Setting out a vision for the city region economy until the year of 2041, it states:

> Our aim is to become the data capital of Europe. We will ensure the social and economic benefits of this capability extend to all. All sections of the community will have the opportunity to develop the key skills that will help to

> end inequalities in our society. We will protect our environment and make best use of our extraordinary built and natural assets, ensuring that the Edinburgh and South East Scotland City Region delivers a just transition to a net zero economy. Our institutions, ancient and modern, will deliver positive outcomes that enhance our local, national and international reputation. (The City of Edinburgh Council, 2021, p. ii)

This vision demonstrates the lasting impact that agents such as the GoF have had on shaping urban-regional futures. In a first step, then, it is important to recognise the university for what it is: a key institution in the politics of urban-regional future-making. The University of Edinburgh is both ancient *and* modern. As a result, it links and shapes both desirable pasts and futures. Imagining the Edinburgh city region as the DCoE is intimately tied to imagining the university as an entrepreneurial institution. As mentioned in Section 4.4., however, entrepreneurial visions clash with more traditional and civic notions of the university. This observation inevitably leads to a critical reflection on the cartography of the university: Does Edinburgh University constitute a satellite in a global system of innovation or a geo-networked planet with strong ties to locality?

In what follows, I contend that the "institutional eclecticism", performed by the University of Edinburgh in the context of the CRD, leads to a blurring and/or shifting of lines of accountability.

The obsession with innovation – itself an indeterminate notion of progress – as a panacea for a range of policy issues leaves the university in an abstracted and ungoverned accountability space. On the one hand, key officials strive to "build a concept of one university" (Interview 21, 2020), thereby subordinating divergent imaginaries of higher education under the rubric of evidence-based, outward-facing, policy relevant knowledge. At the same time, these efforts are inherently constrained, precisely by sociopolitical forces such as the university's historical embeddedness, within the city and the nation. This interplay of forces necessarily begs the following question: To whom shall the university be accountable to?

Illustrating this point, it makes sense to return to the interim conclusion drawn in Section 4.4., in which I stressed that it is far from clear whether the University of Edinburgh – via its engagement in data-driven innovation – is actually able to deliver social and economic benefits to *all*. It remains to be seen how TRADE (talent, research, adoption, data, entrepreneurship) objectives such as increasing the number of people with data skills, harmonising datasets or setting up start-up companies can improve the lives of ordinary citizens in the Edinburgh city region. Does the CRD constitute more than a vehicle for the University of Edinburgh to reimagine itself as globally, networked, entrepreneurial institution?

Confronting key decision-makers with the growing influence of Edinburgh University as a key projective agent and indicating potential issues of democratic accountability, they firmly underscore that it is the Joint Committee – the CRD's executive body comprised of the six political leaders, third sector and business representatives as well as one higher education official – that is ultimately accountable to both the people living in the city region and the UK and Scottish governments. Considering the university's impact on urban-regional future-making, however, it strikes me as highly problematic that policymakers such as the GoF are not accountable to an electorate outside of the University Court or the governments. The establishment of the Regional Enterprise Council (REC) – an advisory body representing third sector and business officials – only *after* all the major decisions had been taken, reflects the current accountability landscape in the context of the CRD. It follows that there is an urgency to better reflect the elevation of the university as a regional policymaker in existing governance structures.

To summarise, the reconfiguration of technoscientific and sociopolitical orders along the "closed-open" and "local-global" axes not only results in a shifting understanding of the modern university (from ivory tower to entrepreneurial institution) but also the associated lines of accountability. Competing with institutions that previously were not part of its self-imaginations and being subjected to global reference points of innovation such as Silicon Valley, to some extent, detaches internal as well as local-regional lines of accountability. Since its foundation in 1583, the university was controlled by the city council. The University Act of 1858 transferred control from the city to the University Court. Given the university's increasingly important role in shaping city-regional development, are there more effective accountability structures that go beyond existing arrangements such as the Joint Committee?

5. Conclusion – lost in translation?

Ever since its foundation, the university has balanced its obligations to the city, the nation, the Empire, the international academic community. But when fewer than half of its students come from Scotland, and the once

> close links with local schools have all but dissolved, the community base is perhaps part of the balance most in need of attention. (Anderson, 2003a, p. 207)

These are the final words of a short book entitled *The University of Edinburgh: An Illustrated History*, published in 2003 by Edinburgh University Press, and written by three prominent Edinburgh historians, Michael Lynch, Nicholas Phillipson and Robert D. Anderson. Reflecting on the various self-imaginations of the university almost twenty years later, I come to a twisted conclusion: the purpose of the university remains unclear, but perhaps that is precisely why it continues to be a central institution shaping the Edinburgh city region.

This chapter was inspired by a desire to better understand how the university – understood as a key truth-making institution in 21st century urban-regional development – is conceptualised by various actors involved in shaping the sociotechnical imaginary of the DCoE. Contributing to literature at the intersection of STS, the sociology of higher education and innovation studies, this chapter makes three key claims. First, the emergence of the city region as a relevant spatial unit and the university as a driver of innovation is inextricably linked to state-making processes. As much as the UK and Scottish governments produce new spaces of innovation, it is also those innovation spaces and agents which help reimagine the modern innovation state.

Second, there are three socio-spatial ideal types or university imaginaries persistently mobilised in the context of the CRD: the university as "ivory tower", as "economic agent" and as "civic institution". Locating the University of Edinburgh amongst those three imaginaries, two key observations emerge. Firstly, the university is transitioning from more traditional self-imaginations revolving around the Humboldtian ideal of the ivory tower to more globally, networked imaginations, rooted in the notion of the "entrepreneurial university". This finding is consistent with existing research on higher education, which traces prevailing conceptions of the modern university: from an ivory tower university of the Enlightenment, to the factory university of the neoliberal knowledge economy, to the universal network university of the 21st century (Matthews, 2022). Secondly, this shift from "closed-local" to "open-global" not only implies an expansion of scale and imagination but, most importantly, different modes of scientific knowledge production.

Finally, the University of Edinburgh – given its status as Scotland's premier university – is inevitably forced to accommodate a range of imaginaries. Contrary to accounts, which diagnose institutional isomorphism resulting from macro factors, such as digitisation, the University performs a mode of "institutional eclecticism". The attempt to sustain its role as a globally oriented agent of innovation-led growth *and* a locally rooted civic institution, however, calls into question existing lines of accountability and purpose. The results are ambivalent: the university remains competitive by addressing different geographically dispersed audiences, yet it risks losing touch with its home base in the face of democratic unaccountability and the uncertain nature of its contributions to local prosperity.

Chapter 7: From Simple to "Dataist Modernity" - Technoscience and the Quest for Control

1. Back to the future

> Mark Logan will be a brilliant asset to the Scottish Government, we've already drawn on his expertise in developing the STER report [Logan Review[1]] and enhancing Scotland's tech credentials. Now we are ready to take this to the next level and ensure entrepreneurial thinking and skillsets are enshrined in the curriculum and across the public, private and social enterprise sectors.
> — Economy Secretary Kate Forbes (2022)[2]

On 7 July 2022, Kate Forbes – Scottish Secretary for the Economy – announced the appointment of Mark Logan as Chief Entrepreneur. Acting as senior advisor to the "Start-up Nation Programme" (Brown, 2021), the former COO of Edinburgh-based unicorn "Skyscanner" is expected to instil entrepreneurialism into Scottish policy-making. This appointment underlines the Scottish Gov-

1 The STER report, also known as the Logan Review (Logan, 2020), constitutes an independent review commissioned by the Scottish Government in May 2020. Examining how Scotland's technology sector can contribute to the country's economic recovery following the Covid-19 pandemic, Mark Logan makes recommendations on how to accelerate and stimulate the maturity of Scotland's technology ecosystem.

2 More information can be found here: https://www.gov.scot/news/appointment-of-chief-entrepreneur/

ernment's attempt to capture the Scots' collective imagination as a community fuelled and driven by innovation and entrepreneurship.

It encapsulates the co-productionist nature of technoscientific and sociopolitical orders. The Scottish Government's dream of independence (Scottish Government, 2022a) and Scotland becoming an "entrepreneurial and ambitious nation" (Brown, 2021) represent two mutually constitutive visions. Importantly, the appointment of Mark Logan demonstrates that the fascination with data-driven innovation and entrepreneurship in the Edinburgh city region is not spatially constrained but plays out over a much wider geography. Instead, it epitomises the intimate relationship between regional, national and global innovation landscapes. In other words, Edinburgh's "data revolution" is linked to national and global efforts to transition into data-driven knowledge economies. How did we get here?

This book presents answers to this question. Tracing the emergence of Edinburgh's innovation apparatus from 2018 to 2021, it provides a snapshot in time and space of a specific imaginary with historical, institutional and technological preconditions. Building on the previous three empirical chapters, in this last part of the study, I want to zoom out and think more broadly about the role of science and technology in the politics of future-making. I ask: How do science and knowledge contribute to city-making? What implications can be drawn from the study of projective agency in the Edinburgh city region for the relationship between science and cities in 21st century societies?

Tackling these questions, I argue that higher education institutions (HEIs) such as the University of Edinburgh become key agents in managing contemporary urban-regional complexity. What used to be the master planner in the early 20th century and the planning state in 1960s and 1970s, are now "subpolitical" organisations such as universities that assume – knowingly or unknowingly – responsibility in shaping smart urban-regional futures.

Advancing this line of argumentation, I will first summarise key findings and present analytical starting points emerging from the previous three empirical chapters. Afterwards, I will reflect on the relationship between technoscience and planning by mobilising two historical examples: Le Corbusier's high modernism from the 1920s and Salvador Allende's cybernetic socialism from the 1970s. Building on these two historical reference points, Section 4 will develop the notion of the "neo-scientisation of urban-regional planning" against the background of what Beck, Giddens and Lash (1994) call "reflexive modernisation". In doing so, I contextualise the emergence of Edinburgh's innovation apparatus with earlier attempts to steer society and situate my empirical findings within a contemporary moment I call "dataist modernity".

2. Key findings and analytical starting points

This book studies the emergence of Edinburgh's innovation apparatus and the role of scientific expertise in shaping sociotechnical imaginaries (STIs) of the city region. It provides a snapshot in time and space of *who* gets to imagine *what* kinds of urban-regional futures. Having traced the politics of future-making in the Edinburgh city region from 2018–2021, a range of empirical observations form the basis for future analysis. In what follows, I want to briefly recapitulate key findings emerging from the previous three chapters in order to set the stage for a concluding analysis on the relationship between science, knowledge and cities in 21st century societies.

In Chapter 4, I disentangled Edinburgh's innovation apparatus by way of focusing on three facets of social change: scales, identities and agencies. I argued that the sociotechnical imaginary of transforming the city region into the "Data Capital of Europe" (DCoE) aligns actors, organisations, discourses and geographical spaces and functions as a panacea that promises eternal economic prosperity. Combining an assemblage-based approach to urban futures (Farías, 2011; Färber, 2014, 2019; Farías and Blok, 2017; Davis and Groves, 2019; Schneider and Lösch, 2019) with Jasanoff's and Kim's (2009, 2015) analytical concept of STIs, I made three key observations.

First, three overlapping policy changes allowed policy makers to reimagine the relationship between science, cities, innovation and politics: 1) since the early 2000s, science and innovation were framed as key drivers of economic growth with the state taking an increasingly interventionist approach; 2) the creation of CRDs, from 2011 onwards, has shifted the focus of economic policy from the state to cities and city regions; and 3) since the Future Cities Demonstrator Competition in 2012, cities are increasingly portrayed as sites and drivers of data-driven innovation.

Second, the DCoE imaginary brings local, regional, national and global identities into conversation. It mediates between Edinburgh as a "city of ideas" (Interview 10, 2020), Scotland as a "tech nation" (Scottish Government, 2022b) and globally "travelling imaginaries of innovation" (Pfotenhauer and Jasanoff, 2017b) such as the Silicon Valley model. Third, Edinburgh's innovation apparatus is enacted and performed through both human and non-human "agents of extension". Translation agents such as the "Gang of Four" (GoF) imagine innovation-driven futures by mobilising both urban-regional governance structures and spaces.

Having laid the empirical foundation of this book, Chapter 5 explored the role of the GoF in structuring power relationships and mediating science-politics boundaries. I argued that reconceptualising the Edinburgh city region as a relevant space of innovation and positioning the University of Edinburgh as its key projective agent required a process of apoliticisation. Invoking what Schiølin (2020) calls "future essentialism", scientific actors such as the GoF apoliticise the present by reimagining desirable pasts and producing "an imaginary of a fixed and scripted, indeed, inevitable, future" (Schiølin, 2020, p. 545). Accordingly, they mobilise three modes of apoliticisation: language, location and data science.

In order for Edinburgh's innovation apparatus to become governable, key decision-makers injected "vocabularies of innovation", which "create[d] a 'trading zone' that allow[ed] different stakeholders to communicate" (Bensaude Vincent, 2014, p. 250). Importantly, the supposedly "trans-ideological properties" (Fox, 2007) of innovation were elevated and cumulated in a belief that technological innovations are inherently attractive and inevitably portray the "next big thing" (Interview 7, 2020). The GoF mobilised location in order to apoliticise data-driven innovation. Invoking Edinburgh as a "truth-spot" (Gieryn, 2018), the six innovation hubs act as cathedrals of a new, "dataist modernity" that envisions a "society of entrepreneurs and service users" (Cowley, Joss and Dayot, 2018). The notion of "dataist modernity" will be developed throughout this chapter.

Finally, besides language and location, it is science, more specifically, *data* science that is mobilised. Underlying this governance technique and imaginary that is represented apolitically is a conviction to see data science simply as the description of a hidden layer of reality rather than understanding it as part of the production of reality itself (McQuillan, 2018). The result of this process of apoliticisation is what I call a "politics of certainty", which separates technoscientific progress from political intent. Instead of acknowledging that technoscientific development is imbued with norms and values about how society ought to develop, scientific actors such as the GoF maintain a veil of apoliticism by clearly separating science, innovation and politics.

Chapter 6 ascended the analytical ladder and examined HEIs as key institutions in 21st century urban development. Identifying three socio-spatial ideal types or university imaginaries – "ivory tower", "economic agent" and "civic institution" – I argued that the University of Edinburgh's shift from traditional self-imaginations towards more globally, networked imaginaries implies a reconfiguration of technoscientific and sociopolitical orders. Putting forward this line of argument, I first acknowledged the co-productionist nature of state (re)scaling and urban governance (re)structuring processes (Brenner, 2004, 2019). That is, as much as the UK and Scottish governments produce new spaces of innovation, such as the city region, innovation spaces and agents such as Edinburgh University also help reimagine the modern innovation state.

Having established this crucial point, I located the university among the aforementioned ideal types and argued that it performs what I call "institutional eclecticism". The latter, defined as the accommodation and interplay of various organisational imaginaries, results in the university being torn between transitioning into an entrepreneurial, networked and global institution on the one hand while remaining locally embedded as a citizen of the Edinburgh city region on the other. As a consequence, I contended, it is crucial to problematise what happens when the University of Edinburgh is elevated from accountability to itself (ivory tower), to accountability to the city region (civic institution), to accountability to globalist reference points of innovation (entrepreneurial university).

In summary, "imagining futures is not innocent" (Hajer and Versteeg, 2019, p. 129). Jasanoff (2004) aptly reminds us that the ways in which we choose to govern knowledge are inseparable from the ways in which we choose to live in society. The emergence of the Edinburgh and South-East Scotland CRD provides a fascinating lens to study fundamental changes in the configuration of technoscientific and sociopolitical orders. Three key messages build the foundation for the forthcoming analysis.

First, there is a power shift in urban-regional governance. City regions are elevated as relevant spaces of innovation and subject to multi-actor governance arrangements. Most strikingly, however, the university as a key agent in the social organisation of knowledge production, dissemination and stabilisation is becoming a broker in driving innovation-based economic growth. Second, following these shifting boundaries between science, politics and society, scientific actors engage in a "politics of certainty" in order to circumvent political responsibility and maintain seemingly clear lines of distinction. Third, some forms of scientific knowledge gain importance in the politics of urban-regional future-making while others are side-lined. More specifically, urban-regional politics increasingly intersects with data science as a relevant form of scientific expertise. This, in turn, calls into question existing ways of *seeing* the city region.

The task ahead is not an easy one. This book tells a story in which several phenomena occur simultaneously. It is showcasing the UK Government's attempt to reignite economic growth by devolving financial and decision-making powers to the subnational level. It is witnessing the Scottish Government's push towards reimagining itself as a tech nation, thereby ensuring politico-economic independence. It is tracing the increasing influence of HEIs in urban-regional development. It is emphasising the enormous impact of globally circulating "best-practice models" such as Silicon Valley and MIT. Finally, it is illuminating an epistemic re-prioritisation towards data science as *the* relevant lens to shape urban-regional life.

How can I possibly do justice to the multiplicity and simultaneity of these phenomena? How can I grasp urban-regional complexity? Rather than sticking to all-encompassing narratives such as neoliberal urbanisation (Venugopal, 2015), I will – keeping the explorative and symmetrical spirit of this research project alive – focus the forthcoming analysis around intersections and nuances of science, politics and cities. Thus, I will first reflect on the relationship between scientific management and urban-regional governance by way of mobilising two historical examples. Afterwards, I will move to the present by developing the notions "neo-scientisation of urban-regional planning" and "dataist modernity".

3. Technoscience and planning

How do science and knowledge contribute to city-making? How does the city region, as a sociospatial setting, affect the production of scientific knowledge? Dierig *et al.* (2003) remind us that science and the city co-produce each other. On the one hand, cities can be regarded as "scientific construction[s]" (Dierig *et al.*, 2003, p. 1), shaped by advances in fields such as engineering, hygiene, social research or urban planning. On the other hand, cities and city regions shape the ways in which knowledge is produced, disseminated and stabilised. For instance, Edinburgh's unique landscape and topography as well as its literati societies during the Scottish Enlightenment inspired geologist James Hutton to develop his *Theory of the Earth*. It follows that rather than treating the city as an empty container in which "science" simply unfolds, it is crucial to recognise the inherently co-productionist nature of science and the city from the start.

Building on this fundamental observation, I now want to introduce two historical examples from the 20th century, which illuminate the intricate relationship between science and scientific management, on the one side, and urban-regional governance, on the other: Le Corbusier's *Plan Voisin* and plans for the *City of Three Million Inhabitants*, dating back to the 1920s, as well as Salvador Allende's "Project Cybersyn", which lasted from 1971–1973.

Both cases – on the city and the nation level – demonstrate how technoscience is mobilised in the management of complexity, how it is invoked in the attempt to order society. Both cases present utopian visions of how the city and the nation respectively ought to develop. Finally, both cases reveal important insights for the use and dissemination of techno-scientific knowledge in contemporary urban-regional development.[3] Serving as reference points for the forthcoming analysis, they shed light on how it was possible to imagine new forms of urban-regional governance through the lens of scientific management in the 1920s, via cybernetics in the 1970s up until emerging forms of smart urbanism from the 1990s onwards (Mora, Bolici and Deakin, 2017; Batty, 2018).

3.1. The city as machine

Like few others, Swiss-born French architect and urban planner Le Corbusier (1887–1965) (see Figure 30) shaped modern urbanism (McLeod, 1983; Scott, 1998; Hall, 2014). A founding member of the *Congrès International d'Architecture Moderne* (CIAM) – an association, consisting of the most prominent architects of the time that was dedicated to promoting the "Modern Movement[4]" – Corbusier is nowadays regarded, controversially, as a pioneer of modern architec-

3 Given the limited scope and disciplinary anchoring of this study, the forthcoming histories are inevitably incomplete. Rather than focusing on a whole body of work, the analysis relies predominantly on Le Corbusier's ([1929] 1987) *The City of To-morrow and its planning* and Eden Medina's (2011) *Cybernetic Revolutionaries*.

4 The "modern movement" in architecture refers to an architectural movement, which emerged during the first half of the 20th century. Reflecting broad transformations in Western industrial societies, it embraced new technologies of construction and followed the ideals of functionalism and efficiency (Blake, 1979). Key propagandists included Ludwig Mies van der Rohe, Walter Gropius and Le Corbusier. Founded in 1928, CIAM was at the heart of the movement and "intended to provide a forum where architectural solutions to the housing, circulation and health problems of the modern metropolis could be discussed and promulgated" (Mumford, 1992, p. 391).

ture and city planning. In order to unpack Corbusian city-making and carve out its relevance for this chapter's argument, I want to highlight three key notions: order, scientific management and technical expertise.

3.1.1. Order

> When man is free, his tendency is towards pure geometry. It is then that he achieves what we call order (Corbusier, [1929] 1987, p. 58)

Corbusier's response to urban complexity was order. Peter Hall (2014) aptly points out that "all Corbusier's ideas need to be seen as a reaction to the city in which he lived and worked from 1916 until shortly before his death in 1965" (p. 239). That city was Paris. Discontent with the chaos of the old Paris, Corbusier urged:

> Paris is a dangerous magma of human beings gathered from every quarter by conquest, growth and immigration; she is the eternal gipsy encampment from all the world's great roads; Paris is the seat of a power and the home of a spirit which could enlighten the world; she digs and hacks through her undergrowth, and out of these evils she is tending towards an ordered system of straight lines and right angles; this reorganization is necessary to her vitality, health and permanence; this clearing process is indispensable to the expression of her spirit, which is fundamentally limpid and beautiful (Corbusier, [1929] 1987, p. 62)

"Straight lines and right angles", "limpid and beautiful". Seeking to tackle congestion and density, Corbusier set out to invent the ideal industrial city. Advancing this agenda, he pushed for geometric simplicity and functional efficiency (see Figures 31 and 32). Importantly, no compromises were to be made with pre-existing cityscapes. Scott (1998) remarks that "the rigor and unity of this ideal city required that it make as few concessions as possible to the history of existing cities" (p. 106). In order to transition from the pre-mechanical into the mechanical age, sweeping change was needed. Reflecting on his "Plan Voisin", which sought to erase and replace large areas of central Paris, Corbusier ([1929] 1987) points out: "my scheme is brutal, because town existence and life itself are brutal: life is pitiless, it must defend itself, hemmed in as it is on all sides by death. To overcome death, constant activity is necessary" (p. 408).

Figure 30: Le Corbusier | Source: Medienhaus Brandenburger[5]

Corbusier's obsession with order and strict functional separation became particularly obvious in his vision for "La Ville Radieuse" – the "radiant city". Building on his utopian plans on "La Ville Contemporaine" and "Plan Voisin", Corbusier – amongst other principles – proclaimed "the death of the street". Loathing what Richard Sennett (2006) would later call the "open city" – concerning "ambiguous edges", "incomplete form" and "unresolved narrative" – Corbusier argued for the complete separation of pedestrian traffic from vehicle traffic. He favoured a city that could be operated top down – a city that is ruled by the plan and embraces the "advances" of the machine-age. The "open city" ought to be replaced by a "city of high-rise towers in a park" (Hall, 2014). Corbusier's solution to density and congested city centres – paradoxically – was to

5 http://medienhaus-brandenburger.de/le-corbusier/

increase density and the amount of green spaces by building high towers on a smaller part of the total ground.

3.1.2. Scientific management

Embracing the "age of the machine", Corbusier's vision for social, urban renewal was heavily shaped by so-called models of scientific management (McLeod, 1983). Inspired by the likes of Frederick Taylor and Henry Ford, Corbusier embraced the factory as an epistemological model for urban governance. Invoking "scientific norms" such as rationality, logic, standardisation and disinterestedness, scientific management was perceived as a way to get rid of politics in the name of reason. McLeod (1983) aptly remarks that "Taylorism [...] offered an escape from ideological conflicts and class divisions: traditional politics would be subsumed by a rational technology of political and economic choice" (pp. 133–134).

Confronted with the devastation of the First World War and inspired by technologies such as the automobile, the airplane and the railway, Corbusier associated Taylorism and mass production as crucial ingredients for social and economic renewal. Laying out his vision for the future city, he proclaimed: "Science has given us the machine. The machine gives us unlimited power. And we in our turn can perform miracles by its means" (Corbusier, [1929] 1987, p. 222). Science and rationality were mobilised in the management of both the city and the factory: "the scientific urban planner is to the design and construction of the city as the entrepreneur-engineer is to the design and construction of the factory" (Scott, 1998, p. 112).

Importantly, underlying this hierarchical and technocratic form of expertise was, what is now referred to as, the "deficit model of science". That is, an imagination of publics as insufficiently knowledgeable. Not only did Corbusier ([1929] 1987) point out that "the public has little notion of what the administration of a great city involves" (p. 167) but sociopolitical dynamics such as elected authorities, parliamentary procedures or public participation were actively dismissed. Instead, what prevailed was "the belief that the human problem of urban design has a unique solution, which an expert can discover and execute" (Scott, 1998, p. 113).

Figure 31: Contemporary City for three million residents | Source: Internet Archive[6]

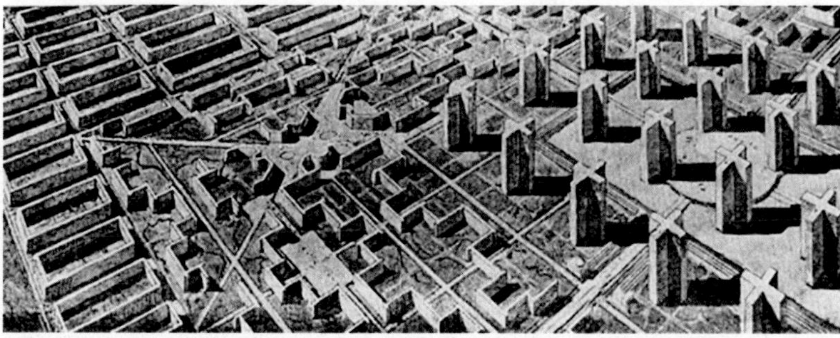

Figure 32: Radiant City | Source: 99% Invisible

This apparent discrepancy between planner and subject was reflected in Corbusier's few real planning achievements on the ground. Generally these were – as Hall (2014) pointedly remarks – "phenomenally unsuccessful" (p. 245). Corbusier's greatest achievement came posthumously in the form of designing the new capital for the Punjab at Chandigarh. India's first planned city after gaining independence in 1947, Chandigarh served as a symbol of progress and modernity (Singh, Kahlon and Chandel, 2019). Taking over from American architect Albert Mayer in 1950 and entrusted by India's first Prime Minister, Jawaharlal Nehru, Corbusier's modernist design was supposed to lead India into the

6 https://archive.org/details/01.18-le-corbusier-la-ville-radieuse-1930/Le-Corbusiers-grandiose-vision-of-the-Ville-Contemporaine-1922-Fondation-Le.png

future. Rather than engaging with prevalent social issues of the Indian population, however, Corbusier was predominantly concerned with visual form and aesthetics. Hall (2014) aptly notes that while the housing was better than what local residents had known before, "the city is heavily segregated by income and civil service rank, recalling La Ville Contemporaine; there are different densities for different social groups, resulting in a planned class segregation" (p. 248).

These fundamental flaws of Corbusier's high-modernist, authoritarian planning were most strongly addressed by American urbanist Jane Jacobs in her seminal work *Life and Death of the American City*. Calling out the master planner's preoccupation with formalism over humanism, aesthetic order over experienced order (Scott, 1998), Jacobs (1961) urges that we should think about cities from the bottom-up. Rather than fighting complexity through clearances and functional segregation, she argues for embracing complexity, informality and unexpectedness. Denouncing Corbusier as "the man with the most dramatic idea of how to get all this anti-city planning right" (Jacobs, 1961, p. 21), Jacobs emphasises epistemic modes of observation and listening while highlighting the organic nature of self-organising communities.

3.1.3. Technical expertise

> I am an architect; no one is going to make a politician of me. Everyone, in his own domain where he is an expert, can apply his special knowledge and carry his solutions to their logical conclusion (Corbusier, [1929] 1987, pp. 411–412)

This remarkable statement by Corbusier appears like the epitome of boundary work. Corbusier perceived of himself as a technical expert devoid of political intentions. In line with the embracing of scientific management models such Taylorism and Fordism, which emphasised science and rationality, Corbusier draws a clear line between his field of expertise – town planning and architecture – and the sociopolitical world he was engaging with. Indicative of his understanding of "scientific planning" are his thoughts on statistics as "the Pegasus of the town planner" (Corbusier, [1929] 1987, p. 166). Upholding a positivist view, and thereby drawing clear lines of distinction between the "science of town planning" and the social world, Corbusier emphasises that "statistics show us the past and foreshadow the future; they provide us with the necessary figures and interpret our graphs and curves" (ibid., pp. 192–193).

It is important to mention, however, that his self-perception as a technical expert applying knowledge did not preclude a deep commitment to societal issues. McLeod (1983) urges us to adopt a more nuanced perspective on Corbusier's politics. Discussing post-modernist critics' and architects' as well as historians' engagement with Corbusier's work, she points out that he "was deeply engaged in social issues, although his involvement generally defies party labels" (p. 132). Rather than being void of specific social and political implications, she argues, Corbusier's vision – deeply rooted in models of scientific management – aimed at linking technology and social change. Contrary to Marxism, Taylorism suggests that social justice is a product of technical rationalisation, not of material equality.

In line with this technocratic interpretation of expertise is Corbusier's fascination for authoritarian control. Celebrating individuals such as Louis XIV, Napoleon I and Georges-Eugène Haussmann as heroes of Paris, he demonstrates a preoccupation with strict hierarchy and swift results rather than deliberation and process orientation. Exemplary of this fascination for top-down control is an image of Louis XIV commanding the building of the Invalides, which concludes Corbusier's *The City of To-morrow and its Planning*:

> Homage to a great town planner. This despot conceived immense projects and realized them. Over all the country his noble works still fill us with admiration. He was capable of saying, "We wish it" or "Such is our pleasure" (Corbusier, [1929] 1987, p. 414)

3.1.4. The state as "planning machine" (Morozov, 2014)

A second historical case that provides important insights for the role of technoscience in co-producing sociopolitical orders represents "Project Cybersyn" ("cybernetics" and "synergy"). Running from 1971 to 1973 – during the presidency of Salvador Allende – this case constituted a complex technological system which was designed to steer Chile's economy into a socialist modernity. Illustrating one of the most ambitious attempts to combine politics and technoscience, "Project Cybersyn" abruptly ended on 11 September 1973 – the day of the violent military coup – which resulted in General Augusto Pinochet taking power and ruling Chile as a dictator until 1990.

Despite its short life, "Project Cybersyn" offers many valuable lessons for the analysis of contemporary attempts to steer society with science and technology. Once again, I want to highlight three notions that are noteworthy – particularly in contrast to Le Corbusier's understanding of planning: control as self-regulation, cybernetic socialism and the "liberty machine".

3.1.5. Control through homeostasis

In order to grasp the essence of "Project Cybersyn", one must understand the intellectual and professional journey of British cybernetician Stafford Beer (see Figure 33). Having already built a lucrative private-sector career at United Steel (at that time, the biggest steel company in Europe), the consulting firm Science in General Management (SIGMA) and the publishing company International Publishing Corporation (IPC), Beer, in July 1971, received an unexpected letter from young Chilean engineer Fernando Flores. Working for newly elected socialist President Salvador Allende,[7] Flores expressed great interest in Beer's cybernetic expertise and asked for his advice on how to apply cybernetics to the management of the increasingly nationalised Chilean economy (Medina, 2011). What followed was the unique confluence of two utopian visions: first, "to implement socialist change peacefully and through existing democratic institutions" and, second, "to build a computer system for real-time economic control more than twenty years before the Internet became a feature of everyday life" (Medina, 2011, p. 3).

Before elaborating in more detail on this form of "cybernetic socialism", it makes sense to briefly introduce the idea of cybernetics and establish Stafford Beer's take on the latter. Deriving from the Greek word *kubernetes*, meaning "steersman" or "pilot", the notion of cybernetics goes back to the American mathematician and MIT Professor Norbert Wiener who initially defined the term as the "control and communication in the animal and the machine" (Wiener, [1948] 2019, p. xxxiv). Emerging in the early 1940s in the context of anti-aircraft defence programs, cybernetics "aimed to establish improved coupling between the human and an ever more complex technological environment"

7 President Salvador Allende was backed by a left-wing political alliance called "Popular Unity". The latter was comprised of most of Chile's left-wing parties, including, among others, the Communist Party and the Socialist Party.

(Picon, 2015, p. 69). Importantly, it was fundamentally concerned with controlling futures. Tracing how cybernetics and communication sciences shaped modes of visioning and reasoning in the second half of the 20th century, Orit Halpern (2014) points out that cybernetics is "a science of control or prediction of future action" and "an endeavour that hopes to tame these future events through the sending of messages" (p. 41).

Firmly rooted in British cybernetics, which – contrary to the US approach – was intimately tied to psychiatry and the "adaptive brain" (Pickering, 2010) rather than military engineering, Beer developed an understanding of control that emphasised self-regulation instead of domination and coercion. Introducing Beer's approach to cybernetics, Medina (2011) remarks that "instead of using science to dominate the outside world, scientists should focus on identifying the equilibrium conditions among subsystems and developing regulators to help the overall system reach its natural state of stability" (p. 28). This natural state of stability is what Beer called "homeostasis"; that is, "the ability of a system to withstand disturbances in its external environment through its own dynamic self-regulation, such as that achieved by coupling subsystems to one another" (ibid.).

"Control through homeostasis" meant finding the right balance between centralised control and individual autonomy. This specific conceptualisation of control was based on what Beer (1972) calls the "Viable System Model" (VSM). This model, if not the central concept in Beer's management cybernetics, presents a management structure that allows agents to regulate "exceedingly complex systems"[8] such as the economy, the brain or the company (Medina, 2011). Modelled after the human brain and nervous system, the VSM represents "a vision of the firm in the image of man" (Pickering, 2010, p. 250). Comprised of five interacting subsystems, the VSM captures different temporalities of organisational action. Subsystems 1–3 are concerned with operations in the present, while systems 4 and 5 are concerned with strategic

8 In his first book *Cybernetics and Management*, Beer (1959) distinguished between three classes of systems: "simple", "complex" and "exceedingly complex" (see Appendix, Figure 41). Simple and complex systems included window catch, penny tossing, the electronic digital computer or stockholding. Crucially, "what those examples have in common, according to Beer, is that they are in principle knowable and predictable, and thus susceptible to the methods of the traditional sciences [physics, statistics]" (Pickering, 2010, p. 222). Exceedingly complex systems such as the economy, the brain or the company, however, were not fully knowable nor predictable and, therefore, subject to the science of cybernetics.

responses to external effects and the overall maintenance of the organisation as a viable entity (Beer, 1972; Pickering, 2010; Medina, 2011).

Underlying this particular vision of cybernetic control more generally, and the VSM more specifically, is what Pickering (2010) calls an "ontology of unknowability" and a "performative epistemology" (p. 25). Contrary to the modern sciences, which advance an ontology of the world as a knowable and representable place, Beer's cybernetics started from the assumption that the world is inherently unknowable and always in a state of becoming. As a result, "the entire task of cybernetics was to figure out how to get along in a world that was not enframable, that could not be subjugated to human designs – how to build machines and construct systems that could adapt performatively to whatever happened to come their way" (Pickering, 2010, p. 32). This insight is crucial since it portrays a fundamentally different understanding of engaging with complexity.

Figure 33: Stafford Beer | Source: InternationalAllende.org[9]

9 https://www.internationalallende.org/antibiografien/stafford-beer/

Beer's model of the "liberty machine", closely related to the VSM, is designed to achieve just that. Functioning as a disseminated network, action is based on information, not authority, and decisions are made in an instant, real-time fashion (Beer, 1971). Section 3.2.3. will pick up the notion of the "liberty machine" by focusing on the one infrastructure that came closest to embodying its real-life manifestation: the infamous operations room (see Figure 34).

3.1.6. Cybernetic socialism

Having laid out some of the basic ideas guiding Beer's notion of cybernetics, it is now time to illuminate how cybernetic science met socialist politics. Key to marrying management cybernetics and democratic socialism was Fernando Flores, freshly appointed general technical manager of the *Corporación de Fomento de la Producción* (CORFO), the State Development Corporation. Responsible for nationalising the Chilean economy, Flores offered Beer the chance to apply his ideas on management cybernetics on a national scale. Crucially, there were deep conceptual similarities between Beer's technoscientific utopia and Popular Unity's – the left-wing political alliance led by Salvador Allende – socialist agenda. Medina (2011) aptly summarises as follows:

> Both Beer and Popular Unity were interested in making structural changes to existing organizations. Both were interested in finding ways to make these changes happen quickly while maintaining the stability of the overall organization. And both were interested in the problem of control but eschewed the idea of ruling with an iron fist. Instead, they wanted to find a balance between individual freedom and top-down control, a balance that preserved autonomy but recognized that maintaining the stability of a state or company may require limiting freedom or sacrificing the needs of some to the needs of others (p. 40)

Put differently, "cybernetic curiosity and political necessity" (Medina, 2011, p. 41) allowed both Beer and Allende to imagine fundamental social, political and economic change. Key to materialising this techno-politico utopia was what Medina (2011) calls the "battle for production". That is, raising industrial production levels became a key mantra for achieving political change. As a result of Allende's nationalisation policy, which led to rapidly increasing numbers of industries and employees under state control, it was all the more important to find an efficient system of managing the economy.

Those involved in constructing "Project Cybersyn", however, faced extraordinary political and engineering challenges. Not only was there enormous time pressure concerning the implementation schedule but, most strikingly, the US government – dismayed by Chile's move towards socialism – initiated an economic blockade against the country while also supporting activities of the political opposition. In other words, there was both a lack of resources in the process of designing "Project Cybersyn" as well as rising domestic political instability.

Despite these extraordinarily difficult circumstances, Beer, Allende and Flores tirelessly pushed "Project Cybersyn" forward. While possessing very limited computer resources – Chile had access to four mainframe machines and, in 1970, only owned approximately 50 computers (Medina, 2011) – Beer made advantage of the fact that the entire computing infrastructure was state-owned. Rather than engaging in complex private-public power struggles and negotiating with a range of different computing centres, Beer only needed to secure support from one agency, the National Computer Corporation. Furthermore, the system Beer imagined fundamentally broke with the previous government's conceptualisation of how to use computers. Rather than seeing the computer as a calculator, he envisioned the computer as a real-time communication network that would allow the Chilean government to make instant decisions on managing economic productivity.

What followed was the design of a complex computing network. Initially called "Project Cyberstride", "Project Cybersyn" consisted of four modules: 1) a telex network (Cybernet), which allowed enterprises to send production data to the National Computer Corporation; 2) a statistical software (Cyberstride) that helped monitor factory performance; 3) an economic simulator (CHECO), which would permit planning and experimenting beyond day-to-day operations; and, lastly, 4) the operations room, modelled after Beer's "liberty machine". The latter – undoubtedly the most iconic symbol of "Project Cybersyn" – shall be discussed in the next section.

3.1.7. The "liberty machine"

> If science is to offer society yet another tool, let us learn from dire experience and build right into the package adequate warnings against its ignorant abuse (Beer, 1971, p. 347)

In his 1971 essay on the "liberty machine", Beer argues that the science of cybernetics can help rethink and redesign existing forms of control and organisation. Lamenting top-down tyranny, he contends that the "liberty machine" – a sociotechnical system that emphasises interconnection, information and instantaneousness – would achieve the right balance between centralised control and individual freedom. Of the four core elements that composed "Project Cybersyn", the operations room, became the real-life manifestation of Beer's "liberty machine".

Located in central Santiago in the interior courtyard of a building occupied by the National Telecommunications Enterprise (ENTEL), the prototype operations room "offered a new image of Chilean modernity under socialism, a futuristic environment for control that meshed with other, simultaneous efforts to create a material culture that Chileans could call their own" (Medina, 2011, p. 115). Importantly, the operations room – created by Ulm School-trained German designer Gui Bonsiepe – reflected the attempt to marry socialism and design. Positioned in a hexagonal space were seven white fibreglass swivel chairs with orange cushions and modern screens on the walls. There were no tables and paper was explicitly banned from the room. The rationale behind this design decision was simple. Limiting the number of chairs to seven would avoid tied votes and prohibiting paper and writing would ease the process of communication.

In addition, particular emphasis was paid to making the room as inclusive as possible. For instance, the armrests of the chairs – comprised of large, geometrical buttons – were designed in such a way that worker participation was possible. With little instruction and no experience needed in using tools such as keyboards, the designers vehemently tried to "prevent it from being a 'sanctum sanctorum for a government elite'" (Medina, 2011, p. 127). Curiously, the construction of the operations room also initiated new working relationships.

For example, the designers around Bonsiepe were looking for a swivel mechanism for the seven fibreglass chairs. Since this particular mechanism was not manufactured in Chile and the government – because of economic sanctions – was running low on cash and supplies, the designers had to become creative. Instead of importing the swivel mechanism, they consulted with local metal workers and, in the end, used an alternative mechanism that allowed the upper part of the chair to move without friction. One cannot help but notice that designing socialism brought along some innovation.

However, for all their attempts to inject socialism into the design of technology those Cybersyn engineers were confronted with complex sociotechnical realities on the ground. This became most obvious during the factory modelling process. Visiting enterprises across the country, the engineers were tasked with building models of production based on key production indicators and their values. It soon became clear that there was an "inconsistent relationship between technology and politics" (Medina, 2011, p. 128). While framing the system as apolitical and scientific to the political opposition, and therefore "neutral", the modelling process turned out to be top down and technocratic. Rather than involving workers in the process, Cybersyn engineers reproduced Taylorist management practices which had disempowered workers in the past. This example demonstrates just how difficult it was to live up to Beer's ideals of the "liberty machine". Ultimately, "Project Cybersyn's" utopianism and scientism could not connect with the struggles that characterised everyday life on the factory floor.

3.1.8. Moving forward

Both – Corbusier's high modernism in the 1920s and Allende's cybernetic socialism in the 1970s – illustrate quests for order. Both cases represent utopian dreams for a new form of governance. Also, both cases – albeit pursuing different political projects – mobilise technoscience in order to cope with uncertainty. Importantly, both cases – arguably – failed in establishing change. Le Corbusier's modern urbanism barely materialised on the ground (Hall, 2014) and Allende's socialist agenda stopped dead in its tracks with Pinochet's coup d'état in 1973 (Medina, 2011).

Nevertheless, I contend that they offer important insights for the role of science and knowledge in contemporary urban planning. Not only do they allow us to trace different configurations of science and planning throughout the 20th century – scientific management in the 1920s, cybernetics in the 1970s – but, most importantly, their residues can be found in today's forms of smart urbanism. In what follows, I want to link Corbusier and Allende with the futures laid out in the context of Edinburgh's innovation apparatus. In doing so, I ask: What role does technoscience play in re-centring the social; that is, aligning social orders around new or different understandings of sociality and the social good. How have "scientific tools" of ordering society changed over time? How does the emergence of big data and artificial intelligence – key ingredients for imagining "smart" futures in the Edinburgh city region – relate to Corbusier's scientific management and Beer's management cybernetics?

Figure 34: The operations room | Source: 99% Invisible[10]

4. The neo-scientisation of urban-regional planning

The two historical examples presented above set the stage for examining how technoscientific and sociopolitical orders co-produce contemporary imaginaries of the *good* society. Ultimately, both cases demonstrate attempts to manage uncertainty through the mobilisation of technoscience. An examination of the relationship between knowledge, science and cities, fundamentally, leads

10 https://99percentinvisible.org/episode/project-cybersyn/

to questioning the relationship between modernity and temporality: How do urban actors – with the help of technoscience – (re)imagine pasts and formulate desirable futures in order to deal with the present?

In what follows, I want to develop the notion of the "neo-scientisation of urban-regional planning" against the background of what Beck, Giddens and Lash (1994) call "reflexive modernisation". I argue that HEIs such as the University of Edinburgh become key agents in managing urban-regional complexity. What used to be the master planner in the early 20th century, the planning state in 1970s, are now "sub-political" organisations such as universities that assume – knowingly or unknowingly – responsibility in shaping smart urban futures. In order to put forward this line of argumentation, I will first elaborate on the notion of "reflexive modernisation" (Beck, 1992; Beck, Giddens and Lash, 1994) before situating my empirical findings in a contemporary moment that I describe as "dataist modernity".

4.1. Simple and reflexive modernity

One epistemological prism that allows us to grasp the interplay of science, knowledge and urban-regional planning represents the notion of "reflexive modernisation". Originally put forward by Ulrich Beck ([1986] 1992) in his seminal work *Risk Society*, "reflexive modernisation" seeks to describe what happens when modernity is confronted with the process of modernisation itself. Contrary to industrial society, modern *risk society* is "concerned no longer exclusively with making nature useful, or with releasing mankind from traditional constraints, but also and essentially with problems resulting from techno-economic development itself" (Beck, 1992, p. 19).

Put differently, the process of modernisation calls into question the very foundations upon which simple, industrial societies are built. Central to the idea of reflexive modernity is the concept of risk. The latter denotes "a systematic way of dealing with hazards and insecurities induced and introduced by modernization itself" (ibid., p. 20). Examples of such hazards and insecurities include climate change, radioactivity and pandemics. It follows that the modernisation of modern society throws coordinates of post-war society – the nation state, the nuclear family and Enlightenment science – into flux. Beck (1992) aptly notes: "Industrial society systematically produces its own endangerment and a questioning of itself through the multiplication and the economic exploitation of hazard" (p. 57).

One hallmark of Western modernity that plays a central role in both (re)producing industrial society and is, subsequently, confronted with making sense of and dealing with the risks produced by it, constitutes science. Reflecting on the politics of scientific knowledge production in the risk society, Beck (1992) introduces the notion of the "demonopolisation of science". That is, science becomes a victim of its own success. Complimentary to industrial and reflexive modernity, Beck identifies two waves of scientisation: primary and reflexive. The former designating a state in which science is "applied to a given world", thereby upholding demarcations such as nature vs. society and experts vs. lay people. The latter phase – reflexive scientisation – denoting a state of complete scientisation "which also extends scientific skepticism to the inherent foundations and external consequences of science itself" (Beck, 1992, p. 155).

Paradoxically, then, science demystifies itself as a result of confronting its own achievements. The Manhattan Project – a research and development endeavour which led to the development of the atomic bomb – clarifies this point. While widely credited as contributing to the end of the Second World War, the risk of a nuclear attack also shaped urban planning and, amongst other factors, led to the dispersal and disintegration of American cities (Light, 2003). Put differently, as much as America's atomic scientists were hailed as innovators, the same innovations produced novel risks, calling into question science's monopoly on truth claims. As a result, risk societies are highly dependent on specialised scientific knowledge to assess and solve complex issues, while scepticism and alienation towards the same scientific knowledges and experts is increasing. Beck (1992) succinctly points out that when rationalism turns on itself, knowledge is transmitted by "agents of knowledge in their combination and opposition, their foundations, their claims, their mistakes, their irrationalities, their truth and in the impossibility of their knowing the knowledge they lay claim to" (p. 55).

One observation that is particularly relevant for the forthcoming argument is what Beck (1992) describes as a "decentring" or "unbinding" of politics. Studying societal changes of the 1970s, Beck (1992) notes that "the waning of welfare state interventionism due to its success and the waves of large-scale technological innovation with as yet unknown future hazards" lead to a reversal of "politics" and "sub-politics". That is, "the potential for structuring society migrates from the political system into the sub-political system of scientific, technological and economic modernisation" (ibid., p. 186). The Covid-19 pandemic demonstrates a case in point. Rather than elected politicians, it was pharmaceutical companies and scientists – first and foremost, public health experts and epidemiologists – who structured Western societies' policy responses to the virus.

Beck's observation that societal steering powers are shifting has implications beyond extreme cases such as the Covid-19 pandemic. It is also reflected in urban modernisation processes. The structuring of urban-regional futures, I argue, is taking place indirectly and often unrecognisably in *scientific* "sites of hyperprojectivity". The latter are "communicative settings, somewhat removed from the flow of day-to-day activity, in which the explicit purpose of talk is to locate problems, visualize alternative pathways, and consider their consequences and desirability" (Mische, 2014, p. 447). Similar to Chile in the early 1970s, where "laboratories and drawing boards [served] as spaces for revolution" (Medina, 2011, p. 65), it is now executive suites, innovation hubs or pubs in Southern France where structures of a new society are imagined. It is these decision-making arenas that emerge as a result of sub-politicisation.

All of sudden, then, it is scientific actors such as the GoF which are "allowed to appear on the stage of social design" (Beck, Giddens and Lash, 1994, p. 22). Crucially, however, and as examined in Chapters 5 and 6, those scientific actors involved in shaping the "DCoE vision" avoid any form of reflexivity in order to keep up the veil of objectivity. Instead they invoke what Scott Lash (1994) calls "formulaic truths" and Enlightenment ideals which suggest that increasing knowledge producing capacities lead to greater control and certainty. Rather than engaging with the risks and uncertainties resulting from increasing knowledge production, they work hard to maintain strict boundaries between science and politics. How is that possible? In what follows, I want to go one step further and analyse how the idea of "data" shapes the relationship between modernity and temporality, science and politics, and knowledge and governance.

4.2. Moving towards a "dataist" modernity

In order to develop the notion of the "dataist modernity", it makes sense to go back in time and contextualise the two historical examples presented in Section 3. What kinds of modernity are Corbusier's high modernism and Allende's cybernetic socialism embodying? How do they feature in the present age of smart urbanism? In what follows, I want to bring empirical reality and theoretical interpretation into conversation. I will first study the three empirical cases (Corbusier, Allende, Edinburgh) in comparison (see Figure 35) and, second, elaborate on the relationship between "neo-cybernetic governance" and "new urban science".

4.2.1. Changing dreams of control

Each of the three empirical cases reveals different technoscience-politics configurations. One comparative dimension that provides a helpful starting point refers to the kind of technoscience mobilised in these respective planning efforts. Envisioning the future city, Corbusier mobilised the "science of town planning" – fundamentally rooted in Taylorist scientific management – to cater for the most dominant technology of the 20th century: the automobile. Functioning as an instrument of the imagination, the automobile allowed Corbusier to envision a social utopia that fundamentally broke with the pre-mechanical age. Jacobs (1961) aptly notes that Corbusier "attempted to make planning for the automobile an integral part of his scheme, and this was, in the 1920's and early 1930's, a new, exciting idea" (p. 23).

Firmly rooted in what Beck *et al.* (1994) call simple or industrial modernity, science is understood to be "applied" to the social world. Invoking traditional scientific norms such as objectivity, neutrality and disinterestedness, Corbusier perceives himself as an expert who simply applies his specialist knowledge to the solution of urban issues such as congestion and density. The use and dissemination of scientific knowledge unfolds in a way that not only draws clear boundaries between science and politics but also conceives of publics as insufficiently knowledgeable, and, therefore, incapable of participating in the design process.

Allende's cybernetic socialism, in contrast, reflects a different configuration of technoscientific and sociopolitical orders. Invoking the science of cybernetics, technoscientific and political objectives were intimately tied together. Building on the computer as a key communication technology, Allende, Flores and Beer laid out a social vision that sought to empower Chilean workers by injecting existing technology with socialist norms and values. Mobilising Beer's management cybernetics in the steering of the economy, "Project Cybersyn" can be considered an early form of algorithmic governance (Medina, 2015). Importantly, it showcases an attempt to combine science, politics and technology in a bottom-up fashion.

Contrary to Corbusier's high modernism, which mobilised science as a tool to dominate from the top down, Chilean cybernetics fed into a form of control that was based on self-regulation. Rather than relying on existing human authority structures, it envisioned the machine – the information processing and communication enabling computer – as a tool that would allow society to break with hierarchical governance. As mentioned in Section 3.2.1., Beer's management cybernetics was based on an ontology of unknowability and becoming. Instead of assuming that technoscience can help dominate and enframe the sociotechnical realities on the ground, "cybernetics staged an ontology in which the fundamental entities [e.g., the economy] were dynamic systems evolving and becoming in unpredictable ways" (Pickering, 2010, p. 31).

The Allende case can be read against the background of reflexive modernity. First and foremost, in the nation state's reliance on political sub-systems such as science and technology in steering society. Put differently, it demonstrates an attempt to mobilise an emerging technology – the computer – for the structuring of the economy. Crucially, this case shows the thin line between acknowledging the politicisation of science and technology while also upholding simple-modernity understandings of science. On the one hand, Beer and his engineers consciously designed socialist norms into infrastructures such as the operations room. On the other, they sought to gain support from members of the political opposition by framing the technological system as apolitical and technocratic (Medina, 2011). It follows that Salvador Allende's cybernetic socialism can be read as a prime example of the clash of industrial and reflexive modernity and its respective modes of scientisation.

Finally, Edinburgh's dream of transforming into the DCoE illustrates a different science-politics configuration. One, which I would locate within a mode of modernity that could be called "dataist modernity". The latter is rooted in the ideology of "dataism" (Brooks, 2013; van Dijck, 2014; Harari, 2017; Fourcade and Gordon, 2020; Davies, 2022). Davies (2022) provides the following helpful definition:

> Dataism is an overestimation of data's affordances that suggests data has its own transformative properties: that if more data is produced and made accessible, and more people are taught to analyse it, then more ambitions will be realised. (p. 148)

Building on these existing conceptualisation efforts, "dataist modernity" shall be defined as a form of modernisation which is rooted in an ontology of knowability and overestimates the power of data in solving complex social problems. Invoking data science as the most relevant source of scientific knowledge (see Chapter 5, Section 4.3.3.), technologies such as big data and artificial intelligence are mobilised in the pursuit of creating data-driven urban-regional futures. What used to be, as Jane Jacobs (1961) pointedly describes, the "highway men" of the 1920s, are now the scientific "data men" of the 2020s. Steeped in the belief that technological innovation will inevitably benefit the city region, scientific actors such as the GoF are embracing a "neo-cybernetic" mode of governance. This includes a *return* to "formulaic truths" (Beck, Giddens and Lash, 1994) and an attempt to dehumanise and apoliticise scientific knowledge production.

The notion of "neo-cybernetic governance" needs unpacking. Inspired by what Picon (2015) terms "neocybernetic temptation" – "the temptation to bring all the problems that face city administrators and decision-makers down to a question of management" (p. 69) – this form of governance combines elements from Corbusier's high modernism and Allende's technological socialism. It brings together Corbusier's vehement apoliticism with Beer's dream of information-based steering power. In order to circumvent what Beck (1992) coins "reflexive scientisation", agents such as the GoF deliberately apoliticise shifting urban governance arrangements. The growing importance and responsibility of universities in regional development, as knowledge transmitters (teaching and research) *and* innovation agents (data infrastructure providers) is externalised through the mobilisation of data science. That is, rather than acknowledging the world-making nature of data science, big data is conceptualised as yet another scientific method that reveals hidden layers of society. In what follows, I want to further unpack the notion of "dataist modernity" by reflecting on the emergence of "new urban science", thereby specifying "Edinburgh's neo-scientisation".

4.2.2. New urban science or the return to certainty?

In addition to studying how cities and city regions are *governed* and *imagined*, it is crucial to take a close look at how they are *known*. What does it mean to study the city through the lens of data science? What kinds of urban futures does data science produce? Karvonen (2020) observes that "there is palpable enthusiasm to increase urban knowledge through the application of big data, ubiquitous sensing, geospatial and social network analyses, algorithms, machine learning, and artificial intelligence" (p. 418). This enthusiasm for knowing the city more intimately through the application of digital technologies is reflected in the emergence of what urban scholars call a "new urban science" (Batty, 2012; Townsend, 2015; Kitchin, 2016; Acuto, Parnell and Seto, 2018).

Figure 35: Modernisation and urban-regional planning | Source: Own illustration, building on Beck, Giddens and Lash (1994)

	Simple modernity	Reflexive modernity	Dataist modernity
Scientific knowledge	Scientific management	Cybernetics	Data science
Technology	Automobile	Computer	Big data and AI
Nature of Science	Scientific rationality	Organised irresponsibility	Organised irresponsibility
Relationship with politics	Demarcation	Sub-politicisation	Apoliticisation
Mode of control	Top-down	Self-regulation	Neo-cybernetic
Empirical example	Le Corbusier's high modernism	Allende's cybernetic socialism	Edinburgh's innovation apparatus

Crucially, then, and keeping the three empirical cases mentioned above in mind, I argue that the rise of a "dataist modernity" does not simply involve a commonly shared fascination for technological innovation but also a social *re*organisation of urban knowledge production. Put differently, it is not just "travelling imaginaries of innovation" (Pfotenhauer and Jasanoff, 2017b) or best-practice models, as found in Silicon Valley, but also "travelling epistemic frameworks" such as "new urban science" which shape the development of smart city regions. Prominent research centres such as MIT's Senseable City Lab[11], UCL's Bartlett Centre for Advanced Spatial Analysis[12] or NYU's Centre for Urban Science and Progress[13] function as epistemic beacons promising to grasp urban complexity through "better theory, better models, and better data" (Townsend, 2015, p. 207).

Crucially, "new urban science" builds the epistemic foundation of what I call "neo-cybernetic governance". Based on greater wealth of urban-regional data, increasing computing power and more mature science, new urban science vows to study and sense *the city as a whole*. Similar to Patrick Geddes' (1904) synoptic view, materialised in Edinburgh's Outlook Tower, new urban science

11 More information can be found here: https://senseable.mit.edu

12 More information can be found here: https://www.ucl.ac.uk/bartlett/casa/bartlett-centre-advanced-spatial-analysis

13 More information can be found here: https://cusp.nyu.edu

promises a holistic, almost Corbusian form of scientific control from above. Acuto *et al.* (2018) urge that academia needs to "forge new knowledge that responds to complex urban challenges and [...] accelerate uptake of urban science by practitioners" (p. 2). Edinburgh's scientific entrepreneurs seem to follow suit. After all, it was the realisation that the university does not own or control its data which motivated the design and implementation of the DDI Programme and its subsequent integration into the CRD (Interview 29, 2020).

Underlying the intimate relationship between "new urban science" and "neo-cybernetic governance" is a recognition that the politics and the science of the future are closely intertwined (Jasanoff, 2020; Wenger *et al.*, 2020). The demand for policy-relevant work and future-oriented knowledge brings science and politics into close contact. This co-constitutive dynamic is clearly reflected in increasingly data-driven approaches to urban governance. Confronted with severe fiscal constraints, rising expectations about service delivery and growing abundance of data and computing power, city administrations – particularly following the financial crisis of 2007/2008 – have started to adopt data-driven approaches in order to rationalise city management.[14] The growing institutionalisation of this new form of governance, in turn, generates increasing demand for data-driven urban analysis (Townsend, 2015). It follows that research centres such as MIT's Senseable City Lab directly respond to policy demands by supplying future-oriented, computationally based urban insights.

Interestingly, underlying this mode of *knowing and sensing the city* in real-time is a healthy degree of apoliticism, which becomes most obvious by distinguishing between the categories of "solution" and "causation". Asked about the relationship between causation and correlation, Michael Flowers, former Chief Analytics Officer of the City of New York, famously pointed out: "I am not interested in causation except as it speaks to action. Causation is for other people, and frankly it is very dicey when you start talking about causation. [...] I can't dick around, frankly, thinking about other things like causation right now" (quoted in Mayer-Schönberger and Cukier, 2013, p. 189). Hooked by the appeal

14 One prominent example constitutes the creation of New York City's Mayor's Office of Data Analysis, which was established in 2013 under the leadership of Michael Bloomberg. Led by Michael Flowers, the first Chief Analytics Officer and Chief Open Platform Office of the City of New York, the Office developed a fire risk assessment tool based on data concerning over 900,000 buildings (Flowers, 2013).

of cybernetic control, bureaucrats like Flowers "can be effective while remaining ignorant of the causal mechanisms at play" (Morozov, 2014, p. 4). This insight is crucial and points to the "externalisation of responsibility", mentioned in Section 4.2. Rather than engaging with underlying sociopolitical dynamics, such as inequality and poverty, policymakers such as Flowers or the GoF focus on the "now", seeking to make existing governing processes more efficient and avoiding any responsibility in (re)producing inequalities.

The notion of the "externalisation of responsibility" reveals strong similarities with what Beck (1998) terms "organised irresponsibility". The latter "helps to explain how and why the institutions of modern society must unavoidably acknowledge the reality of catastrophe while simultaneously denying its existence, hiding its origins and precluding compensation or control" (Beck, 1998, p. 18). Viewing the interplay of "new urban science" and "neo-cybernetic governance" through the lens of "organised irresponsibility", I contend that big data represents the latest mechanism of disavowing responsibility in urban-regional policymaking. Following this line of thought, one is confronted with a paradox: more knowledge and more data implies more complexity; yet more data (supposedly) suggests a greater control of complexity.

Scientific institutions such as Edinburgh University and actors like the GoF, however, circumvent Beck's mode of reflexive scientisation by making an epistemological move: instead of viewing big data as a tool that helps map *uncertainty*, thereby self-reflexively acknowledging a potential loss of control, they invoke data as a tool to map *risk*, thus suggesting calculability and control. To put it another way, *mapping risk* upholds "neo-cybernetic governance". *Mapping uncertainty* fundamentally challenges "neo-cybernetic governance". Neo-cybernetic governance pertains to a society characterised by fear of what technoscience brings next.

Beck's (2009) distinction between three types of "future insecurity" – threat, risk and manufactured uncertainty – helps clarify this point. The notion of "threat" pertains to pre-modern societies' engagement with the future. Threats such as natural disasters are conceived as externalisable, non-human and attributable to God or nature. "Risk" and "uncertainty", in turn, are categories of modern and reflexive societies. Risks denote "calculable uncertainties" – futures that are "made knowledgeable by measurement" (Beck, 2009, p. 296). Here, human agents engage actively with unknown futures in order to form a basis for rational decision-making.

"Manufactured uncertainties", in contrast, are "dependent on human decisions, created by society itself [...] incalculable, uncontrollable and in the final analysis no longer (privately) insurable" (ibid., p. 293). Applying this threefold distinction to the concept of "neo-cybernetic governance", it appears that scientific actors like the GoF promote a form of scientific management that denies uncertainty. Rather than self-reflexively confronting the uncertainties involved in scientific knowledge production – thereby recognising the need for a rethinking of the role of universities in steering society – they hide behind the veil of dataist scientificity. Instead of acknowledging the inherent contingencies involved in the interplay of knowledge, facts and values, they invoke big data as the saviour of scientific authority in 21st century society. "Dataist modernity" thus implies first and foremost an "overestimation of the power of data" (Brooks, 2013).

The epistemological reversal of risk and uncertainty contains a more fundamental ontological assertion. Cities and city regions are understood as systems of control, made up of "flows of information and feedback" (De Monchaux, 2011). Asked about why it is that data-driven innovation has come to play such a significant role for imagining the future city region, this senior architect of the DDI Programme remarks:

> "I think we have come ... the people that have looked at it I think have come to realise that cities are ultimately about two types of things. They are about movement and transaction. And movement and transaction involve data, they involve choices. So the choices I make about how I move around a city inform both how I experience it but inform providers of services to provide services in different ways. And the way I transact with the city so the choices I make about what I buy or don't buy or, have a significant impact. And that's not just about cities, that's just modern life." (Interview 10, 2020)

One cannot help but notice the Corbusian undertone. Urban-regional governance is imagined from the top down. Underlying this particular ontology is what Cowley *et al.* (2018) call a "service-user modality of publicness" where urban-regional residents are primarily imagined as consumers of urban services such as public transport, water or broadband infrastructure. Big data and smart technologies, in turn, promise to enhance service provider-service user interactions "by providing more accurate or 'real-time' data for decision-making, new possibilities for users to contribute to services through automated or voluntary feedback, and new channels of accountability" (Cowley *et al.*, 2018, p. 63).

Two materialisations on the ground signify Edinburgh's enthusiasm for neo-cybernetic governance. One is the Edinburgh International Data Facility[15] (EIDF), which is supposed to provide the underlying data infrastructure for the DDI Programme. The other is Edinburgh City Council's (2020b) recent announcement to extend its partnership with the Canadian IT and business consulting firm CGI to become "one of the world's smartest capital cities".[16] At the heart of these efforts lie ambitions to establish a smart city operations centre, which is supposed to "deliver transformative digital services using the likes of AI, the 'Internet of Things' and Advanced Analytics" (ibid., p. 1). It appears that the lure of cybernetic modernity has made its way from Chile to Edinburgh.

4.3. Neo-scientisation?

What role does science and knowledge play in imagining, knowing and governing the contemporary city region? Is it possible to speak of a "neo-scientisation of urban-regional planning"? The previous comparison and contextualisation of Corbusier's high modernism, Allende's cybernetic socialism and Edinburgh's emerging innovation apparatus against the background of Beck *et al.*'s (1994) "reflexive modernisation", allows me to draw some preliminary conclusions. Each case reveals different attempts to mobilise science and technology in the steering of society – be it on the city, city region or nation level. To speak of a "*neo*-scientisation" of planning, however, one must pinpoint exactly what it is that characterises the contemporary moment. *Who* scientises? *What kind* of science? *How* does it unfold?

The short answer to these questions is that sub-political actors such as HEIs are increasingly engaging in a "neo-cybernetic" form of governance by mobilising data science as an authoritative source of urban-regional knowledge. In order to disentangle this claim, it makes sense to go back in time. Similar to Thomas Kuhn's (1962) scientific "paradigm shifts", Peter Hall (2014) identifies major changes in planning theory during the 20th century. He notes that

> "the subject [of planning] changed from a kind of craft, based on personal knowledge and a rudimentary collection of concepts about the city, into an

---

15 More information can be found here: https://ddi.ac.uk/about-us/eidf/

16 More information can be found here: https://www.edinburgh.gov.uk/news/article/12962/council-extends-partnership-with-cgi-as-it-aims-to-become-one-of-the-world-s-smartest-cities-

> apparently scientific activity in which vast amounts of precise information were garnered and processed in such a way that the planner could device very sensitive systems of guidance and control, the effects of which could be monitored and if necessary modified" (p. 437).

The previous journey from Le Corbusier in the 1920s to Allende's Chile in the 1970s to Edinburgh's innovation apparatus in the 2020s, reflects this shift from the master planner – supposedly free from political intervention – to a mode of planning that finds itself increasingly confronted with the political. During this period, planning – the attempt to understand and manage uncertainty (Abbott, 2005) – became increasingly scientised. Moving from Corbusier's (1987) fascination for statistics as "the Pegasus of the town planner" (p. 166) to Beer's dream of cybernetic control until the GoF's enthusiasm for data science emerged, urban-regional planning is deeply imbued with science and technology. From being geared towards a static world, with stagnant populations and slow economic growth, to being confronted with a dynamic and accelerated world, shaped by rapidly growing urban economies, science and technology have been persistently mobilised as instruments of the imagination.

Crucially, however, significant differences arise pertaining to the specific role that technoscience played and still plays in shaping visions of the future. Comparing Allende's cybernetic socialism and Edinburgh's emerging innovation apparatus a fundamental divergence exists in the way that technoscience is mobilised. In the Chilean case, technoscience represents a tool that acts *in support of* broader policy shifts. Here, cybernetics was mobilised in order to implement Allende's socialist agenda. In Edinburgh's case, technoscience functions as *an end in and of itself*. Instead of embedding data-driven innovation in norms and values about how the city region ought to develop, technologies such as big data and AI are framed in an apolitical manner. Scientific actors like the GoF are "prisoners of the present" – they lack imagination beyond the now. This erratic, almost arbitrary, form of scientisation is fuelled by data determinism and a lack of political vision.

Now, one can of course counter this claim by pointing to the apolitical self-perception of scientific agents: How can we possibly expect political imagination from individuals who conceive of themselves as exclusively scientific? It is here where the "politics of certainty" (i.e., the strict separation of techno-scientific progress from political intent) enters the scene. As long as scientific actors such as the University of Edinburgh uphold this mode of "primary scientisation" (Beck, 1992), thereby denying the co-productionist nature of science and politics, technoscience will remain disembodied from societal governance. Building on Beck's (1992) three scenarios of a possible future – "reindustrialisation", "democratisation of technological transformation" and "differential politics" (p. 224) –, I wonder if "dataist modernity" simply implies a return to the modalities of "industrial society" but with a pinch of data?

Studying actors like the GoF, what dominates is a sense of Scottish Enlightenment 2.0 – unswerving faith in progress. The risks of the 21st century are assumed to be met by the creative forces of scientific and technological development. Invoking ideas such as "Edinburgh's data revolution", key decision-makers – paraphrasing Beck (1992) – respond to the question "What should we do?" as follows: "Data will tell us what to do". What prevails is an "as if" mode of engagement with the future. Scientific actors like the GoF act *as if* the future is knowledgeable. They act *as if* data science held a lease on urban-regional truths. The result is a mode of neo-scientisation that is stuck in the past.

5. Conclusion – entering "dataist modernity"?

This chapter dealt with the impossible task of presenting an overall discussion that brings the main components of this book – science, knowledge, cities, imaginaries, innovation, politics – into conversation. In an effort to comprehensively illustrate the relationship between science and urban-regional development, I have drawn on two historical examples: Le Corbusier's high modernism from the 1920s and Salvador Allende's cybernetic socialism from the 1970s. Comparing the emergence of Edinburgh's innovation apparatus with the planning utopias of the 1920s and 1970s enabled me to examine how technoscience is mobilised in the context of managing complexity – how it is drawn upon in the attempt to order society.

In doing so, I examined the three empirical cases, all of which represent a different interplay of technoscientific and sociopolitical orders, against the background of what Beck, Lash and Giddens (1994) termed "reflexive modernisation". By situating the science-politics configurations of Corbusier and Allende in "simple" and "reflexive" modernity, respectively, I identified contemporary planning efforts, such as Edinburgh's vision to transform itself into the DCoE, in a moment I described as "dataist modernity" which denotes a form of modernisation which is rooted in an ontology of knowability and overestimates the power of data in solving complex social problems.

Rooted in the ideology of "dataism" (Brooks, 2013; van Dijck, 2014; Davies, 2022), "dataist modernity" is best captured through the notions of "neo-cybernetic governance" and "organised irresponsibility". The former signifies a mode of control that conceives of the city as a system of flows and feedback, which can be managed centrally and top down. The latter points to scientific actors' dataist dreams to change the status quo while insisting "with the innocent face of objectivity that they are not responsible for any of the issues decided in these plans" (Beck, 1992, p. 224). As a result, we are confronted with a mode of neo-scientisation that resembles a re-orientation to the past rather than a self-reflexive engagement with the future. The notion of "dataist modernity" thus allows us to capture the persistent – rather than changing – nature of techno-science-politics configurations in contemporary urban-regional planning.

Chapter 8: The View after the Questions

1. Change is in the air

Figure 36: "Smart city" Edinburgh | Source: Edinburgh Living Lab[1] "Smart Places"

Change is in the air – quite literally. Edinburgh and the Edinburgh city region are being reimagined. The city region is no longer perceived as just a "city of managers" (Interview 7, 2020), a "city of money" (Perman, 2018) or a city of science, but increasingly envisioned as the "Data Capital of Europe" (DCoE) – a breeding ground for data-driven innovation, AI, and robotics. Figure 36, taken from the Edinburgh Living Lab's "Smart Places" project – captures its

1 More information can be found here: https://edilivinglab.medium.com/smart-places-where-people-data-and-technology-come-together-to-make-better-places-ac77c1fcaf84

essence: an assemblage consisting of Edinburgh Castle, Queensferry Crossing, trees and city dwellers on a bright, sunny day is complemented by the digital. The digital is invoked through symbols such as the cloud, Wi-Fi and a lightbulb hanging above the city region.

Interestingly, the image combines notions of 1950's US suburbia, on the one hand, and data, which appears to fit seamlessly into the natural order, on the other. Codes about the good life – community, domesticity and private family life – are brought into conversation with metaphors such as the cloud, which signals the organic nature of data. Socio-material implications of the cloud, such as the operation of data centres, environmental degradation or precarious working conditions, are neither mentioned nor implied. Instead, codes and values of the past are combined with symbols of the future to shape desirable city-regional imaginaries. Crucially, this image comes with a promise: "add data and innovation to urban-regional life and you will solve complex social issues such as child poverty or inequality while boosting the economy".

This book critically interrogated the social basis of this promise and its limits. Investigating the emergence and early implementation of the DCoE vision, I followed actors and institutions around the city region to gain a better understanding of what it means to inject Edinburgh – a place imbued with tradition and focus on the past – with technologies of the future. In doing so, one key realisation, as mentioned in the previous chapter, was that "imagining futures is not innocent" (Hajer and Versteeg, 2019, p. 129). Technoscientific innovations – whether defined as a "conscious effort to help the urban community to solve the challenges or make the city work better" (Interview 7, 2020) or simply speculative fiction in real life – imply a profound reordering of the self, society and milieu. Langdon Winner (1986) aptly points out that

> [T]echnological innovations are similar to legislative acts or political foundings that establish a framework for public order that will endure over many generations. For that reason, the same careful attention one would give to the rules, roles, and relationships of politics must also be given to such things as the building of highways, the creation of television networks, and the tailoring of seemingly insignificant features on new machines. The issues that divide or unite people in society are settled not only in the institutions and practices of politics proper, but also, and less obviously, in tangible arrangements of steel and concrete, wires and semiconductors, and nuts and bolts (p. 29)

From 2018–2022, I have been exploring the interplay of the tangible and the intangible, science and politics, pasts, presents and futures. In studying Edinburgh's innovation apparatus, I was confronted with the constant attempt by policymakers to herald the dawn of a new modernity while preserving existing power structures. I have experienced what it means for local institutions to meet local, regional and national expectations while ensuring global competitiveness. Finally, I have realised the enormous importance of the past and the urgency of the present when it comes to imagining alternative futures. In concluding this book, I will do three things. First, I am going to summarise the main contributions of this research project. Afterwards, I will reflect on its limitations. Finally, I am going to provide a brief overall reflection on three analytical categories – the actors, the institution, the bigger picture – as well as future research paths.

2. Contributions

This ESRC-funded study has investigated the process by which urban planning agents imagine futures and the sociotechnical conditions which help establish the credibility of their expectations. Aiming to shed light on the practice of city-making, this book has examined how urban planning agents co-produce and diffuse specific urban-regional "sociotechnical imaginaries" (STIs) (Jasanoff and Kim, 2009, 2015). To this end, I have empirically engaged with the implementation of a contemporary urban-regional redevelopment project, the Edinburgh and South-East Scotland City Region Deal (CRD). This deal represents a £1.3bn investment by the Scottish and UK governments and local partners over the next fifteen years, which is designed to accelerate data-driven innovation and inclusive growth. Officially signed in August 2018, the CRD represents a unique case study for at least two reasons. First, it constitutes the first higher education-driven CRD in the UK. Second, it aligns a cluster of considerably powerful actors – political leaders, higher education officials, tech entrepreneurs – around one specific sociotechnical imaginary – establishing the city region as the DCoE.

Conceptually rooted at the intersection of Science and Technology Studies (STS) – particularly its co-productionist strand (Jasanoff, 2004) – and urban studies, in this book, I argued that future-making in the Edinburgh city region is characterised by an unequal distribution of "projective agency" (Emirbayer and Mische, 1998) and driven by an entrepreneurial class at the intersection of science and politics. Shedding light on the complex interplay of expectations and expertise in economic policy-making, this study has demonstrated the growing influence of global higher education institutions (HEIs) such as the University of Edinburgh in shaping desirable futures. More specifically, the concept of "dataist modernity" was developed to characterise a form of modernisation rooted in an ontology of knowability and which overestimates the power of data in solving complex social problems. By crafting this line of argument, this study has extended co-productionist STS on "smart urbanism" to take fuller account of the intricate relationship between technoscience and urban-regional planning.

I have advanced four key claims. First, there is a power shift in urban-regional governance. City regions are elevated as relevant spaces of innovation and subject to multi-actor governance arrangements. Most strikingly, however, the university as a key agent in the social organisation of knowledge production, dissemination and stabilisation is becoming a broker in driving innovation-based economic growth. Second, following these shifting boundaries between science, politics and society, scientific actors engage in a "politics of certainty" in order to circumvent political responsibility and maintain seemingly clear lines of distinction between the political and the apolitical. Third, some forms of scientific knowledge gain importance in the politics of urban-regional future-making while others are side-lined. More specifically, urban-regional politics is increasingly intersecting with data science as a relevant form of scientific expertise. This, in turn, calls into question existing ways of *seeing* the city region. Fourth, the relationship between technoscience and urban-regional planning can be conceptualised against the background of what Beck, Giddens and Lash (1994) call "reflexive modernisation" and captured in a mode of modernity that I call "dataist modernity".

Building on these findings, this book makes several contributions to existing bodies of academic literature in STS, sociology and urban studies that examine the interplay of temporality, innovation and urban-regional development. First and foremost, this book contributes to research at the intersection of co-productionist STS and urban studies by interpreting cities as complex sociotechnical assemblages, and conceptualising urban innovation through Jasanoff's and Kim's (2015) notion of STIs. Contrary to existing accounts, which tend to focus either on the state or on corporate actors as central sites for the evolution and expansion of STIs, this research project has emphasised the role of cities, more generally, and HEIs, in particular, as important drivers of organisational action in their own right when it comes to actualising smart city visions. Thus, this book not only puts forward a multi-scalar understanding of innovation processes but also sheds light on universities as central knowledge-producing *and* society-shaping actors.

In addition, this study adds to the smart city discourse in three ways. Firstly, it provides an empirical account of smart urbanism in the Edinburgh city region – a city region that – to this date – has received only limited academic attention (Karvonen, Cugurullo and Caprotti, 2019; Currie, Knox and McGregor, 2022). Secondly, it points to the important role of HEIs in driving urban-regional economic development. Much of the existing literature has focused on tech corporations as central agents in smart city initiatives (Vanolo, 2014; Wiig, 2015; Sadowski and Bendor, 2018). This study sheds light on the University of Edinburgh as a major force behind data-driven urbanism. Building on existing accounts in urban studies, which suggest the increasingly important role of universities in the co-production of smart cities (Karvonen and van Heur, 2014; Karvonen, Martin and Evans, 2018), I have demonstrated that data-driven innovation functions as a vehicle for the university to collect urban-regional data and, at the same time, enhance its institutional reputation.

Thirdly, this book contributes to existing accounts in the sociology of higher education and innovation policy studies. Recent academic work demonstrates how the role of universities evolved from being a resource for policy and economy to being an engine of economic growth (Etzkowitz and Leydesdorff, 1995; Berman, 2012; Thorp and Goldstein, 2013; Audretsch, 2014). This study goes one step further by suggesting that global HEIs such as the University of Edinburgh constitute more than just economic actors. Building on recent studies at the intersection of STS and innovation policy (Pfotenhauer and Jasanoff, 2017b; Pfotenhauer and Juhl, 2017; Pfotenhauer *et al.*, 2022), which suggest that techno-scientific and socio-political orders are always co-produced, I have argued that shifting self-imaginaries of the modern university call into question existing lines of political accountability. Put differently, it is paramount to acknowledge that the sites and objects of politics are no longer simply those of formal state institutions but also lie in the hands of increasingly important players, such as HEIs, that fall outside the bounds of meaningful political control.

Finally, this book provides a preliminary theoretical framework of the interplay of technoscience and urban-regional development. Comparing the emergence of Edinburgh's innovation apparatus with two planning utopias from the 1920s and 1970s – Le Corbusier's high modernism and Salvador Allende's cybernetic socialism – I have situated my empirical findings against the background of what Beck, Giddens and Lash (1994) term "reflexive modernisation". Doing so, this study identifies contemporary planning efforts such as Edinburgh's vision to transform itself into the DCoE in a moment it coins as "dataist modernity". That is, a form of modernisation which is rooted in an ontology of knowability and which overestimates the power of data in solving complex social problems.

Generally rooted in the ideology of "dataism", which implies an overestimation of the power of data (Brooks, 2013; van Dijck, 2014; Fourcade and Gordon, 2020; Davies, 2022) – "dataist modernity" is best captured through the notions of "neo-cybernetic governance" and "organised irresponsibility". The former signifying a mode of control that conceives of the city as a system of flows and feedback, which can be managed centrally and top down. The latter pointing to scientific actors' dataist dreams to change the status quo while insisting "with the innocent face of objectivity that they are not responsible for any of the issues decided in these plans" (Beck, 1992, p. 224). As a result, the current mode of neo-scientisation is characterised by an "ontology of knowability" (Pickering, 2010) that is oriented towards the past instead of embracing the uncalculability of the uncertain future.

3. Limitations

In attempting to capture the emergence and early implementation of Edinburgh's innovation apparatus as comprehensively and in as much detail as possible, this study inevitably reveals limitations. Most strikingly, the exclusive – but not unconscious – focus on the Edinburgh city region underrepresents data-driven urban-regional reforms in other regions of the UK, Europe and the wider world. Given this partial perspective, I am accountable to both the insights this research project has generated as well as its inevitable blind spots. Given the limited scope of this book, I have chosen to focus explicitly on the characteristics of Edinburgh's emerging innovation apparatus, challenging conventional notions of techno-scientific urbanisation such as neoliberalism, institutional isomorphism, and the "practice turn" in innovation policy (Venugopal, 2015; Pfotenhauer and Jasanoff, 2017b; Barrioluengo, Uyarra and Kitagawa, 2019).

Nevertheless, future research could build on the results of this study and take a comparative view, thereby offering a more systematic review of smart city-regional development. For instance, it would be highly intriguing to conduct a systematic comparison of the role of data-driven innovation across all of the city region and growth deals in the UK. In addition, it might be instructive to look more closely at powerful HEIs, such as Oxbridge, Manchester, Edinburgh and some of the London institutions, to examine the intricate relationships between land ownership, income flows and city-regional development.

Besides empirical and methodological omissions, key concepts such as "apoliticisation" or "dataist modernity" remain underdeveloped. This shortcoming is also primarily due to the limited scope of this study. Nevertheless, it is crucial to further flesh out these ideas in order to not only empirically advance research at the intersection of STS and urban studies, but to promote theory building more broadly. After all, I have viewed the PhD as an intellectual playground that has allowed me to think big and bold. Future research could build on the back of the conceptual framework developed in Chapter 7, for instance, to further substantiate efforts to historicise the relationship between technoscience and urban-regional planning. Crucially, rather than setting out to document change, the notion of "dataist modernity" allows us to study stability and stabilisation since it illuminates the persistence of forms of knowledge injection in urban-regional planning efforts. For instance, additional empirical work on the Edinburgh Futures Institute – a key site and institution mediating Edinburgh's pasts and futures – would help shed light on how scientific knowledge production and urban development relate to each other.

Finally, this research project has examined a "moving target". As the empirical focus was on a three-year period (2018–2020), key events such as the Covid-19 pandemic or overall political changes in the UK and Scotland are not sufficiently taken into account in the analysis of Edinburgh's innovation apparatus. Once again, the limited duration of my investigation prevented a more detailed analysis. Future research should thoroughly examine how "external" shocks such as the Covid-19 pandemic shape the politics of urban-regional imagination.

4. Reflection

What happens if we take seriously the notion that the ways in which knowledge is produced, stabilised and disseminated cannot be separated from the ways in which we choose to live in society? What role does data-driven innovation play in re-centring the social? That is, aligning social orders around new or different understandings of sociality and the social good. Tracking the emergence and early implementation of Edinburgh's innovation apparatus, this book has told the story of powerful individuals, such as the "Gang of Four" (GoF) and increasingly important urban-regional stakeholders such as the University of Edinburgh, falling prey to and performing what can be called "dataist modernity". To conclude this study, I want to briefly revisit three units of analysis from Chapter 5–7 – the actors, the institution and the bigger picture – in order to draw some meta-level conclusions about the relationship between science and cities.

4.1. The actors

This book has shown that some actors are granted permission to engage in the politics of future-making while others are left outside. More specifically, it has highlighted the city-shaping role of scientific actors – most prominently, data scientists (Chapter 5). What are the implications of recognising the increasing influence of data scientists in shaping our cities and city regions? If we accept that the "scientific data men" of the 2020s exert world-shaping power beyond their traditional fields of expertise, can we conceptualise them as *political* actors?

Before engaging with this question, it is important to briefly comment on the "left behind" – those disenfranchised from imagining futures. In Chapter 4, I introduced the notion of the "social stratification of futurity" in order to emphasise different levels of "projective agency" (Emirbayer and Mische, 1998). Social stratification of futurity denotes the degree to which economic actors are exposed to the escalating nature of capitalist dynamics, and the extent to which their decisions are influenced by futurity. This definition entails at least three key claims. First, human experience is centrally mediated by time (Emirbayer and Mische, 1998; Heidegger, 2010; Skotnicki and Nielsen, 2021). Second, expectations and narratives function as central devices that help decision-makers deal with the uncertain nature of the future (Brown and Michael, 2003; Beckert, 2016). Finally, engagement with and capacity to shape futures varies because projective agency is unevenly distributed (Bourdieu, 1979; Davis and Groves, 2019; Hajer and Versteeg, 2019).

This study has demonstrated that the politics of future-making in the Edinburgh city region is driven by an elite at the intersection of science and politics. To put it bluntly, the idea of turning the city region into the DCoE originated as a backroom deal and has remained – until now – a project for the already privileged. Asked if the "woman on the street" is aware and can relate to this particular urban-regional vision, this high-ranking third sector representative responded as follows:

> "But if you've never had to fill out a 26-page universal credit form on Wi-Fi, sitting outside the supermarket because you're actually trying to piggyback their café Wi-Fi, because you don't really have the money to sit inside and have a cup of tea, then you will never know what that feels like.
>
> "So, no, the data capital of the world in terms of Edinburgh does not make sense for a huge swathe of our population. And if we're going to do something [...] we as data capital and data region then then actually we need to think dramatically about the infrastructure of that data, not just the amazing potential in terms of ... I don't know, space travel or solving vaccine conundrums or any of those things" (Interview 37, 2020)

This remarkable statement sums up existing discrepancies between policy makers' envisioned "landscapes of innovation" (Zukin, 2020d) and the socio-material realities on the ground. It underlines not only that "projective agency" constitutes a privilege of certain, exclusive social strata (Bourdieu, 1979; Atkinson, 2013), but more importantly that futures imagined with data and technology are anything but self-evident for most people. Most importantly, this statement highlights the urgent need to grant ordinary citizens a "right to the smart city" (Kitchin, Cardullo and Di Feliciantonio, 2019). This means developing alternative conceptualisations of citizenship and adopting more sustainable, collective and normative visions of the smart city. A promising starting point – currently implemented in Barcelona – represents the concept of "technological sovereignty", which denotes "citizens' capacity to have a say and participate in how the technological infrastructure around them operates and what ends it serves" (Morozov and Bria, 2018, p. 22).

Acknowledging the powerful and examining the relationship between data science, ethics and politics, Green (2021) argues that "data scientists must recognize themselves as political actors engaged in normative constructions of society and evaluate their work according to its downstream impacts on people's lives" (p. 249). Throughout this research project, I have showcased the "boundary work" (Gieryn, 1983) involved in denying the co-constitutive processes of the "politicisation of science" and the "scientisation of politics" (Weingart, 1999). Actors like the GoF artificially limit their self-imagination, and thus maintain a position from which they can separate techno-scientific progress from political intention.

Crucially, however, "the scientific workplace functions as a key site for the production of social and political order" (Jasanoff, 2012, p. 264). Green (2021) aptly points out that the invocation of neutrality and apoliticism itself constitutes a political move "because neutrality does not mean value-free – it means acquiescence to dominant social and political values, freezing the status quo in place" (p. 253). But what would happen if key decision-makers openly and self-reflectively acknowledged that transforming the Edinburgh city-region into the DCoE meant not only mobilising data and innovation to drive economic growth, but also reconceptualising the cornerstones of city-regional society more broadly?

So far, this hope remains a pipe dream. Instead, we are confronted with city-regional technocrats obsessed with the idea that data-driven innovation constitutes a panacea for the most complex social problems. This book makes a case for the closer study of what Kitchin *et al.* (2017) call "smart city epistemic community", referring to the "network of knowledge and policy experts that share a worldview and a common set of normative beliefs, values and practices with respect to addressing urban issues, and work to help decision-makers identify and deploy technological solutions to solve city problems" (p. 1). More specifically, it urges scholars to examine higher education officials' role in shaping urban-regional planning: Who are the individuals within the university that are driving the reimagination of organisational identities? What definition of "social good" do they put forward? How do variables such as gender and intellectual socialisation influence the latter?

4.2. The institution

Global HEIs such as the University of Edinburgh matter. As key truth-making infrastructures of the 21st century, they sit at the nexus of knowledge and power. They educate large proportions of the future workforce, host a wide range of research activities, own real estate across city regions, employ thousands of people, make their expertise available to wider sections of society and preserve knowledges of the past (Arbo and Benneworth, 2007; Berman, 2012a; Goddard *et al.*, 2014; Sørensen and Traweek, 2021). This research project has traced the increasingly influential role of universities in urban-regional innovation economies. It has shown that the transition from more traditional self-imaginations, revolving around the Humboldtian ideal of the "ivory tower", to more globally and networked imaginations, rooted in the notion of the "entrepreneurial university", calls into question existing lines of accountability. Accountability landscapes become increasingly complex when global education institutions such as the University of Edinburgh are shifting from accountability to itself (ivory tower), to accountability to the city region (civic institution), and to accountability to globalist reference points of innovation (entrepreneurial university).

If we accept that modern institutions of knowledge such as the university increasingly perform what I have called a mode of "institutional eclecticism" and imagine themselves as allotropic (i.e., as existing in many different states at once), to whom are they accountable? Currently, institutions such as Edinburgh University cater for plural constituencies and values – ranging from education, markets and consumption to civil society. Depending on the audience – local residents, national governments, technologists, global competitors – HEIs invoke different understandings of the self. If we take seriously the idea that different imaginaries of the university are representative of specific ways of producing and valuing knowledge, then it is crucial to unmask these respective configurations of the knowledge-power dyad.

I wonder – to paraphrase Jasanoff (2003) – if one can reimagine universities as "institutions of humility" in order to develop a more constructive idea of the relationship between science and society. Building on Beck's (1992) analysis of the "risk society", Jasanoff (2003b) argues for a different relationship between scientific expertise, policy-making and the public – one that moves beyond the "speaking truth to power" model. Instead, she introduces the concept of "technologies of humility", which denote "methods, or better yet institutionalized habits of thought, that try to come to grips with the ragged fringes of human understanding – the unknown, the uncertain, the ambiguous, and the uncontrollable" (Jasanoff, 2003b, p. 227). Humility refers to a mode of knowledge-making that recognises the inherent limits of prediction and control and calls for novel forms of public accountability. Rather than rendering issues more technical and "scientific", expertise is thought of as opening up debate and creating space for reflection.

Speaking of "institutions of humility" thus provokes the following questions: What would happen if global HEIs like the University of Edinburgh left the "spaceship" (Interview 40, 2020) and told public stories that explained how and why science is the way it is? What would happen if they told public stories that acknowledged the contingency of knowledge production thereby challenging the politics of certainty? What would happen if universities developed a working definition of inclusive, data-driven urbanism that went beyond cosmetic concepts such as "data ethics"?

Unfortunately, the idea of powerful knowledge institutions self-reflexively acknowledging the inherent contingencies of knowledge production currently seems as utopian as the proposal to politicise the "scientific data men" of the 2020s. After all, one of, if not the greatest strength of powerful institutions is to provide a sense of certainty. Nonetheless, and particularly from a STS perspective, more scholarly work needs to be done on why it is that city-regional actors agree – rather than disagree – on specific data-driven futures. Rather than naively calling for the democratisation of higher education governance, it will be crucial to further unpack the construction of shared perceptions of urban futures.

Unmasking the politics of certainty seems to be one way of opening up a political space in which novel forms of accountability can be negotiated. Importantly, this could trigger a rethinking or reimagination of the notion of democracy and democratisation itself. Just as the notion of dataist modernity embodies the poverty of urban imagination, I contend that a simple call for democratisation represents a lack of political imagination. In other words, what would happen if the creation of novel accountability spaces included the reimagination of the "we" – against the background of a healthy scepticism towards the idea of democracy as the default mode of social order?

4.3. The bigger picture

An essential part of what I have described as the "politics of certainty" constitutes what Pickering (2010) calls "ontology of knowability" – "the idea that the world is finitely knowable" (p. 390). The latter is intimately linked to the epistemological reversal of risk and uncertainty. That is, big data and data science are seen as tools that help map risk rather than uncertainty, transforming cities and the city region into systems of control rather than spaces of exploration. Kitchin (2014b) aptly remarks that "the drive towards managing and regulating the city via information and analytic systems promotes a technocratic mode of urban governance which presumes that all aspects of a city can be measured and monitored and treated as technical problems which can be addressed through technical solutions" (p. 9). The statement below captures the gist of this particular ontology. When asked about the role of uncertainty in city-regional planning, this former council leader responded as follows:

> "Edinburgh is working on the basis that the growth we are experiencing right now, the growth of the population and economic growth continues. That is our basic assumption. That is the assumption that is trusted upon by the administration. [...] Uncertainty isn't a huge part of our thinking. It probably should be a bit more but then it is quite hard to quantify it when you're looking at dealing with growth and success rather than shrinking and failure." (Interview 1, 2019)

Globally interconnected crises such as the Covid-19 pandemic highlight the radical contingencies involved in making futures actionable. Rather than assuming sustained growth and success, wouldn't it be productive to adopt an ontology of unknowability? What would happen if city-regional decision-makers actively acknowledged what Braun (2015) calls the "futurity paradox"? That is, "the inevitable tension between the future-oriented nature of economic action and the 'uncalculability' of the uncertain future" (p. 368).

Reconceptualising the relationship between science and cities means reimagining urban-regional ontology and epistemology. Challenging existing practices and assumptions in smart city initiatives and urban science, Kitchin (2016) argues for a "re-orientation of how the city is conceived [...], a re-casting of the epistemology of urban science [...] and the adoption of ethical principles" (p. 11). Global HEIs such as Edinburgh University are uniquely positioned to drive ontological, epistemological and ethical change. Moving forward, I want to make one empirical and one conceptual claim: first, universities must provide better access to future-imagining practices and, second, there needs to be more research into the role of technology in expectation formation.

Regarding the first demand, institutions such as Edinburgh University sit at the nexus of knowledge and power, the local and the global, pasts, presents and futures. To ensure a just transition from data-*informed* to data-*driven* cities and city regions, it is crucial to acknowledge that "future-imagining practices have future-making consequences" (Lindner and Meissner, 2019, p. 17). This book has shown that the emergence of Edinburgh's innovation apparatus was characterised by a lack of public debate and a general discrepancy between key decision-makers visions and local residents' lived realities. In order to live up to the notion of *inclusive* growth, the *Regional Prosperity Framework* (The City of Edinburgh Council, 2021) – the successor of the CRD – must go beyond lip service.

That means that scientific actors actively shape the creation of alternative perspectives. Hajer and Versteeg (2019) notice that "there is an uneven battle for the minds in which academics focus their best attention on their own internal communication via peer-reviewed outlets, while business and industry dominate the public debate with teleological predictions of a coming future of smart cities and self-driving vehicles" (p. 129). Rather than reproducing data-deterministic urban-regional visions, powerful universities such Edinburgh must ensure access to future-imagining practices and develop alternative imaginaries of the social good (Zehner, 2022).

Second, to further substantiate the concept of "dataist modernity", it would be highly interesting to examine the influence of technologies such as AI, robotics and big data centres on expectation formation. Svetlova (2022) emphasises that "models, algorithms and various technological devices [...] aim to compensate for agents' bounded rationality and, hence, co-determine and profoundly change the information agents search for, how they proceed with the collected data and which futures they imagine" (p. 842). In Chapter 7, I have termed the prevalent mode of control in "dataist modernity" as "neo-cybernetic". There is a need to further explore not only how increasing computing power and more sophisticated data analytics produce ontological claims of the "real-time city" (Kitchin, 2014b), but also how these technologies are shaping the ways in which human agents imagine urban-regional futures.

One way in which futurity is built into existing infrastructures of envisioning represents Edinburgh's supercomputing centre.[2] In a recent episode of the "Data Capital" podcast, Lee Rand – EMEA Sales Director at Hewlett Packard Enterprise and one of the architects behind Edinburgh's high performance computing efforts – points out that "with the new world of data analytics and data science, you can't afford to throw away any data".[3] It is precisely this "wholesale erosion of data minimisation principles [...] i.e., the undermining of purpose specification and use limitation principles" (Kitchin, 2016, p. 10) that characterises the relationship between big data and planning. In other words, the possibility of collecting and analysing large amounts of urban-regional data changes human agents' engagement with the future. How does this horizon of constant potentiality shape the politics of urban-regional futures?

2 More information can be found here: https://www.epcc.ed.ac.uk

3 More information can be found here: https://open.spotify.com/episode/054gLDjRrY5R40ktQcU7Az

Figure 37: "The View after the Questions" | Source: Ade Adesina, Royal Scottish Academy[4] *2018*

I want to conclude this book with Nigerian-born and Scotland-based artist Ade Adesina[5] and his work "The View after the Questions" (see Figure 37). This piece of art represents a commentary about global connections, which addresses the interrelationships between the economy, the environment, development, and the survival of endangered species. In his linocut work, Adesina evokes urban utopias and dystopias, urging people to reflect on the past, the present and the future. By depicting the infinite complexity and multi-layeredness of urban life, Adesina sensitises people to a world that is not dominated by people, science and technology.

---

4 More information can be found here: https://www.royalscottishacademy.org/artists/331-ade-adesina-rsa/works/2646-ade-adesina-rsa-the-view-after-the-questions-2018/

5 Ade Adesina was born in Nigeria in 1980 and studied printmaking at Gray's School of Art, Robert Gordon University, Aberdeen. Adesina's work has been exhibited worldwide and he is a member of several prestigious associations, including the Royal Scottish Academy, the Royal Glasgow Institute of The Fine Art, the London Group and is an associate member of the Royal Society of Painter's & Printmakers. Adesina's work focuses on the effects of climate change, energy consumption and deforestation, combining objects and places from his African heritage with British culture. More information can be found here: https://www.royalscottishacademy.org/artists/331-ade-adesina-rsa/

"The View after the Questions" is a plea against simplification. It is a plea against the separation of man and nature, science and politics, past and future. Most importantly, it conveys a non-modern ontology of unknowability and becoming (Pickering, 2010). Contrary to the image used at the beginning of this chapter, it decentres human agency and conveys a sense of humility. What looms above the city is not data but a horizon of environmental degradation. Symbols such as the hourglass, a tree of life and flying sharks allude to fragility and the constant possibility that our world is literally being turned upside down. In this sense, Adesina takes up one of the central tensions of this book; namely, people's attempt to control the world in the face of the fact that it is uncontrollable.

So what is the view after the questions? The answer is simple: more questions! What happens if we shift away from technoscience being the central component in improving urban-regional life? What role can self-reflexive forms of science and technology play in supporting broader policy change? Is it possible to imagine urban-regional futures that are not primarily based on technoscientific innovations? Perhaps the goal should not be to reclaim or repair existing forms of smart urbanism but to come up with an entirely different conceptualisation of the "smart city".

Bibliography

Abbott, J. (2005) 'Understanding and managing the unknown the nature of uncertainty in planning', *Journal of Planning Education and Research*, pp. 237–251.

Abram, S. and Weszkalnys, G. (2011) 'Introduction: Anthropologies of planning—Temporality, imagination, and ethnography', *Focaal*, 2011(61), pp. 3–18.

Acuto, M., Parnell, S. and Seto, K. C. (2018) 'Building a global urban science', *Nature Sustainability*, pp. 2–4.

Adam, B. (1998) *Timescapes of Modernity*, Routledge.

Adam, B. (2010) 'History of the future: Paradoxes and challenges', *Rethinking History*, pp. 361–378.

Adam, B. and Groves, C. (2007) *Future Matters*. BRILL.

Addie, J. P. D., Keil, R. and Olds, K. (2015) 'Beyond town and gown: Universities, territoriality and the mobilization of new urban structures in Canada', *Territory, Politics, Governance*. Taylor & Francis, 3(1), pp. 27–50.

Agamben, G. (2009) 'What is an Appatatus?', *What is an Appatatus? and Other Essays*. Stanford University Press, pp. 1–24.

Aibar, E. and Bijker, W. E. (1997) 'Constructing a city: The Cerdà Plan for the extension of Barcelona', *Science Technology and Human Values*, 22(1), pp. 3–30.

Albino, V., Berardi, U. and Dangelico, R. M. (2015) 'Smart cities: Definitions, dimensions, performance, and initiatives', *Journal of Urban Technology*, 22(1), pp. 1–19.

Alvial Palavicino, C. (2015) 'The Future as Practice A Framework to Understand Anticipation in Science and Technology', *Tecnoscienza*, 6(2), pp. 135–172.

Amin, Ash; Thrift, N. (2002) *Cities. Reimagining the urban*. Polity Press.

Amin, Ash; Thrift, N. (2016) *Seeing like a city*. Wiley.

Amin A. and Graham S. (1997) 'The Ordinary City', *Transactions of the Institute of British Geographers*, 22(4), pp. 411–429.

Anderson, B. (1983) *Imagined Communities – Reflections on the Origin and Spread of Nationalism*. Verso.

Anderson, R. D. (2003a) 'Age of Expansion', in *The University of Edinburgh: an illustrated history*. Edinburgh University Press.

Anderson, R. D. (2003b) 'The Construction of a Modern University', in *The University of Edinburgh: an illustrated history*. Edinburgh University Press.

Anderson, R. D., Lynch, M. and Phillipson, N. (2003) *The University of Edinburgh: An Illustrated History*. Edinburgh University Press.

Angelidou, M. (2015) 'Smart cities: A conjuncture of four forces', *Cities*, 47, pp. 95–106.

Anttiroiko, A., Laine, M. and Lönnqvist, H. (2020) 'City as a Growth Platform: Responses of the Cities of Helsinki Metropolitan Area to Global Digital Economy', pp. 1–21.

Appadurai, A. (1990) 'Disjuncture and Difference in the Global Cultural Economy', *Theory, Culture & Society*, 7(2), pp. 295–310.

Arbo, P. and Benneworth, P. (2007) 'Understanding the regional contribution of higher education institutions: A literature review', *OECD, Directorate for Education, Working Paper N. 9*, (1), pp. 1–78.

Arup (2013) 'Solutions for Cities: An Analysis of the Feasibility Studies from the Future Cities Demonstrator Programme', *Technology Strategy Board*.

Atkinson, W. (2013) 'Class habitus and perception of the future: Recession, employment insecurity and temporality', *British Journal of Sociology*, 64(4), pp. 643–661.

Audit Scotland (2020) *Scotland's City Region and Growth Deals*.

Audretsch, D. B. (2014) 'From the entrepreneurial university to the university for the entrepreneurial society', *Journal of Technology Transfer*, 39(3), pp. 313–321.

Baert, P. and Shipman, A. (2005) 'University under siege? Trust and accountability in the contemporary academy', *European Societies*. European Societies, pp. 157–185.

Baldwin, D. L. (2021) *In the shadow of the ivory tower: how universities are plundering our cities*. Bold Type Books.

Bandola-Gill, J. (2020) 'The legitimacy of experts in policy: navigating technocratic and political accountability in the case of global poverty governance', *Evidence & Policy: A Journal of Research, Debate and Practice*, xx(xx), pp. 1–19.

Bank, B. B. (2020) *New Future Fund to support innovative UK businesses – British Business Bank*.

Bank, W. (2022) *Urban Development Overview*.

Barrioluengo, M., Uyarra, E. and Kitagawa, F. (2019) 'Understanding the evolution of the entrepreneurial university. The case of English Higher Education institutions', *Higher Education Quarterly*, 73(4), pp. 469–495.

Batty, M. (2012) 'Building a science of cities', in *The New Science of Cities*. MIT Press.

Batty, M. (2018) *Inventing Future Cities*. MIT Press.

Beck, U. (1992) *Risk Society: Towards a New Modernity*. SAGE.

Beck, U. (1998) 'Politics of Risk Society', in Franklin, J. (ed.) *The Politics of Risk Society*. Blackwell.

Beck, U. (2009) 'World risk society and manufactured uncertainties', *Iris*, 1(2), pp. 291–299.

Beck, U., Giddens, A. and Lash, S. (1994) *Reflexive Modernization – Politics, Tradition and Aesthetics in the Modern Social Order*. Stanford University Press.

Beck, U. and Wehling, P. (2013) 'The politics of non-knowing: An emerging area of social and political conflict in reflexive modernity', *The Politics of Knowledge*, pp. 33–57.

Beckert, J. (1996) 'What is sociological about economic sociology? Uncertainty and the embeddedness of economic action', *Theory and Society*, 25(6), pp. 803–840.

Beckert, J. (2009) 'The social order of markets', *Theory and Society*, 38(3), pp. 245–269.

Beckert, J. (2013a) 'Capitalism as a System of Expectations: Toward a Sociological Microfoundation of Political Economy', *Politics and Society*, 41(3), pp. 323–350.

Beckert, J. (2013b) 'Imagined futures: Fictional expectations in the economy', *Theory and Society*, 42(3), pp. 219–240.

Beckert, J. (2016) *Imagined futures: Fictional expectations and capitalist dynamics*. Harvard University Press.

Beckert, J. and Bronk, R. (2018) 'Uncertain futures: Imaginaries, narratives, and calculation in the economy', *Uncertain Futures: Imaginaries, Narratives, and Calculation in the Economy*, pp. 1–333.

Beckert, J. and Ergen, T. (2021) 'Transcending History's Heavy Hand: The Future in Economic Action', in *Handbooks of Sociology and Social Research*, pp. 79–94.

Beckert, J. and Suckert, L. (2021) 'The future as a social fact. The analysis of perceptions of the future in sociology', *Poetics*. Elsevier B.V., 84(December 2018), p. 101499.

Beer, S. (1959) 'Cybernetics and Management'. The English Universities Press Ltd.

Beer, S. (1971) 'The liberty machine', *Futures*, 3(4), pp. 338–348.

Beer, S. (1972) *Brain of the Firm: The Managerial Cybernetics of Organization*. Wiley.

Bell, D. (1973) *The Coming of Post-industrial Society: A Venture in Social Forecasting*. Basic Books.

Benjamin, W. (1999) *The Arcades Project*. Harvard University Press.

Bensaude Vincent, B. (2014) 'The politics of buzzwords at the interface of technoscience, market and society: The case of "public engagement in science"', *Public Understanding of Science*, 23(3), pp. 238–253.

Berman, E. P. (2012a) *Creating the Market University: How Academic Science Became an Economic Engine, Creating the Market University: How Academic Science Became an Economic Engine*.

Berman, E. P. (2012b) *The Market University, Creating the Market University*.

Blackmore, J. and Sawers, N. (2015) 'Executive power and scaled-up gender subtexts in Australian entrepreneurial universities', *Gender and Education*, 27(3), pp. 320–337.

Blaikie, N. W. H. and Priest, J. (2017) *Social research: Paradigms in Action*. Wiley.

Blake, P. (1979) 'The Modern Movement: What Went Wrong?', *The Wilson Quarterly*. Winter, 3(1), pp. 121–127.

Blok, Anders; Farias, I. (2016) *Urban cosmopolitics. Agencements, Assemblies, Atmospheres, Handbook of Urban Geography*.

Bloor, D. (1991) 'Knowledge and Social Imaginary'. The University of Chicago Press.

Borup, M. *et al.* (2006) 'The sociology of expectations in science and technology', *Technology Analysis and Strategic Management*, pp. 285–298.

Bourdieu, P. (1979) *Algeria 1960: The Disenchantment of the World, The Sense of Honour, The Kabyle House or The World Reversed*. Cambridge University Press.

Boyer, R. (2018) 'Expectations, Narratives, and Socio-Economic Regimes', in *Uncertain Futures – Imaginaries, Narratives, and Calculation in the Economy*. Oxford University Press.

Brandtner, C. *et al.* (2017) 'Enacting governance through strategy: A comparative study of governance configurations in Sydney and Vienna', *Urban Studies*, 54(5), pp. 1075–1091.

Braun, B. (2015) 'Governing the future: the European Central Bank's expectation management during the Great Moderation', *Economy and Society*, 44(3), pp. 367–391.

Brenner, N. (1998) 'Global cities, glocal states: Global city formation and state territorial restructuring in contemporary Europe', *Review of International Political Economy*, 5(1), pp. 1–37.

Brenner, N. (2004) *New state spaces: urban governances and the rescaling of the statehood*. Oxford University Press.

Brenner, N. (2019) *New Urban Spaces: Urban Theory and the Scale Question*. Oxford University Press.

Brenner, N., Madden, D. J. and Wachsmuth, D. (2011) 'Assemblage urbanism and the challenges of critical urban theory', *City*, 15(2), pp. 225–240.

Bronk, R. (2009) *The Romantic Economist*. Cambridge: Cambridge University Press.

Brooks, D. (2013) *The Philosophy of Data*, *The New York Times*.

Brown, N. and Michael, M. (2003) 'A sociology of expectations: Retrospecting prospects and prospecting retrospects', *Technology Analysis and Strategic Management*, 15(1), pp. 3–18.

Brown, N., Rip, A. and Van Lente, H. (2003) 'Expectations In & About Science and Technology', *Background paper for the ' expectations ' workshop*, (June), pp. 1–14.

Brown, R. (2021) *The Next Start-Up Nation? A Manifesto for an Independent, Entrepreneurial and Ambitious Scotland*.

Burgess, E. W. (1925) *The Growth of the City: An Introduction to a Research Project*.

Callon, M. (1984) 'Some elements of a sociology of translation: domestication of the scallops and the fishermen of St Brieuc Bay', *The Sociological Review*, pp. 196–233.

Callon, M. (1999) 'The Role of Lay People in the Production and Dissemination of Scientific Knowledge', *Science Technology & Society*, 4(1), pp. 81–94.

Callon, M. (2008) 'Economic Markets and the Rise of Interactive Agencements: From Prosthetic Agencies to Habilitated Agencies', in *Living in a Material World: Economic Sociology Meets Science and Technology Studies*.

Calzada, I. and Cobo, C. (2015) 'Unplugging: Deconstructing the smart city', *Journal of Urban Technology*, 22(1), pp. 23–43.

Capello, R., Olechnicka, A. and Gorzelak, G. (2012) 'Universities, cities and regions: Loci for knowledge and innovation creation', in *Universities, Cities and Regions: Loci for Knowledge and Innovation Creation*, pp. 1–376.

Caragliu, A., del Bo, C. and Nijkamp, P. (2011) 'Smart cities in Europe', *Journal of Urban Technology*, 18(2), pp. 65–82.

Carrillo, F. J. *et al.* (2014) *Knowledge and the city: Concepts, applications and trends of knowledgebased urban development, Knowledge and the City: Concepts, Applications and Trends of Knowledge-Based Urban Development*. Routledge.

Castells, M. (1977) *The Urban Question. A Marxist Approach*. Edward Arnold.

Castells, M. (1999) 'Grassrooting the Space of Flows', *Urban Geography*, 20(4), pp. 294–302.

Castells, M. (2010) *The Information Age: Economy, Society and Culture. Volume I. The Rise of the Network Society, Wiley-Blackwell.*

Castoriadis, C. (1998) *The Imaginary Institution of Society*. MIT Press.

Charles, D. (2003) 'Universities and Territorial Development: Reshaping the Regional Role of UK Universities', *Local Economy*, 18(1), pp. 7–20.

Charles, D. (2006) 'Universities as key knowledge infrastructures in regional innovation systems', *Innovation*, 19(1), pp. 117–130.

Charmaz, K. and Belgrave, L. L. (2012) 'Qualitative interviewing and grounded theory analysis', *The SAGE Handbook of Interview Research: The Complexity of the Craft*, pp. 347–366.

Chatterton, P. and Goddard, J. (2003) 'The response of HEIs to regional needs', in *Economic Geography of Higher Education: Knowledge, Infrastructure and Learning Regions*, pp. 19–41.

City Of Edinburgh Council (2018) *Edinburgh and South-East Scotland City Region Deal.*

City Of Edinburgh Council (2019) *Edinburgh and South East Scotland City Region Deal Joint Committee Edinburgh and South East Scotland Regional Growth Framework Item number Executive Summary.*

Clark, B. R. (1998) *Creating entrepreneurial universities: organizational pathways of transformation.*

Coletta, C. *et al.* (2018) *Creating smart cities*. Routledge.

Collins, Harry; Evans, R. (2017) *Why Democracies Need Science*. Polity Press.

Collins, H. and Evans, R. (2007) *Rethinking Expertise*. University of Chicago Press.

Connell, R. (2019) *The Good University: What Universities Actually Do and Why It's Time for Radical Change, ZedBooks.*

Corburn, J. (2005) 'Introduction', in *Street Science: Community Knowledge and Environmental Health Justice*. MIT Press.

Corbusier, L. (1987) *The city of to-morrow and its planning*. Dover.

Cowley, R., Joss, S. and Dayot, Y. (2018) 'The smart city and its publics: insights from across six UK cities', *Urban Research and Practice*, 11(1), pp. 53–77.

Cronon, W. (1991) *Nature's metropolis: Chicago and the Great West*. W.W. Norton & Company.

Currie, M., Knox, J. and McGregor, C. (2022) *Data Justice and the Right to the City*. Edinburgh University Press.

Data Driven Innovation Programme (2022) *Annual Review 2022*.

Datta, A. (2015) 'A 100 smart cities, a 100 utopias', *Dialogues in Human Geography*, 5(1), pp. 49–53.

Davies, H. C. (2022) 'Rescuing Data Literacy from Dataism', in Currie, M., Know, J., and McGregor, C. (eds) *Data Justice and the Right to the City*. Edinburgh University Press.

Davies, J. S. (2002) 'Urban regime theory: A normative-empirical critique', *Journal of Urban Affairs*, 24 1(1), pp. 1–17.

Davis, J. and Groves, C. (2019) 'City/future in the making: Masterplanning London's Olympic legacy as anticipatory assemblage', *Futures*. Elsevier Ltd, 109(January), pp. 13–23.

Degen, M. (2018) 'Timescapes of urban change: The temporalities of regenerated streets', *Sociological Review*, 66(5), pp. 1074–1092.

Department for Business Energy and Industrial Strategy (2021) *UK Innovation Strategy Leading the future by creating it*.

Dequech, D. (1999) 'Expectations and Confidence under Uncertainty', *Journal of Post Keynesian Economics*, 21(3), pp. 415–430.

Dequech, D. (2006) 'The new institutional economics and the theory of behaviour under uncertainty', *Journal of Economic Behavior and Organization*, 59(1), pp. 109–131.

Desmond, M. (2014) 'Relational ethnography', *Theory and Society*. Theor Soc, 43(5), pp. 547–579.

Dewey, J. (1922) *Human Nature and Conduct: An Introduction to Social Psychology*. Henry Holt. A

Dierig, Sven, Lachmund, J. and Mendelsohn, J. A. (2003) 'Introduction: Toward an Urban History of Science', *Osiris*, 18, pp. 1–19.

Dierig, Sven., Lachmund, J. and Mendelsohn, J. A. (2003) *Osiris: Science and the city*. University of Chicago Press.

van Dijck, J. (2014) 'Datafication, dataism and dataveillance: Big data between scientific paradigm and ideology', *Surveillance and Society*, 12(2), pp. 197–208.

DiMaggio, P. J., Powell, W. W. and Di Maggio, P. J. (1983) 'The iron cage revisited: Institutional isomorphism and collective rationality in organizational fields', *American Sociological Review*, 48(2), pp. 147–160.

Dodd, Nigel; Wajcman, J. (2016) 'Simmel and Benjamin', in *The Sociology of Speed: Digital, Organizational, and Social Temporalities*. Oxford University Press.

Dörre, Klaus; Lessenich, Stephan; Rosa, H. (2015) *Sociology, Capitalism, Critique*. Verso.

Van Driessche, R., Ache, P. and Lagendijk, A. (2017) *Dissecting the urban(ized) binoculars. 'Looking at' urban futures*.

Edinburgh, U. of (2020) *Annual Report and Accounts for the Year to 31 July 2020*.

Emirbayer, M. and Mische, A. (1998) 'What Is Agency?', *American Journal of Sociology*, 103(4), pp. 962–1023.

Etzkowitz, H; Leydesdorff, L. (1995) 'The Triple Helix–University-Industry-Government Relations: A Laboratory for Knowledge Based Economic Development', *Easst Review*, 14(1), pp. 14–19.

Etzkowitz, H. (2002) 'The Triple Helix of University – Industry – Government The Triple: Implications for Policy and Evaluation', *Working Paper*, 11, pp. 1–17. A

Etzkowitz, H. (2003) 'Studies of science Etudes sur la science Innovation in innovation×: the Triple Helix of university – industry – government relations', *Social Science Information*, 42(3), pp. 293–337.

Eubanks, V. (2018) 'The Allegheny Algorithm', in *Automating Inequality. How High-Tech Tools Profile, Police and Punish the Poor*.

Evans, James; Karvonen, Andrew; Raven, R. (2016) *The experimental city*, *Science*. Routledge.

Evans, J. and Karvonen, A. (2014) '"Give Me a Laboratory and I Will Lower Your Carbon Footprint!" – Urban Laboratories and the Governance of Low-Carbon Futures', *International Journal of Urban and Regional Research*, 38(2), pp. 413–430.

Eyal, G. (2012) 'Spaces between fields', in *Bourdieu and Historical Analysis*, pp. 1–12.

Eyal, G. (2019) *The crisis of expertise*. Wiley.

Ezrahi, Y. (1990) *The Descent of Icarus: Science and the Transformation of Contemporary Democracy*. Harvard University Press.

Ezrahi, Y. (2012) 'Chapter 2: The Revival and Contemporary Legacy of Giambattista', in *Imagined Democracies: Necessary Political Fictions*, p. 2000.

Fama, E. F. (1965) 'The Behavior of Stock-Market Prices', *The Journal of Business*, 38(1), pp. 34–105.

Farahani, A. (2017) 'Urban Growth Machine', *The Wiley-Blackwell Encyclopedia of Social Theory*, pp. 1–4.

Färber, A. (2014) 'Low-budget Berlin: Towards an understanding of low-budget urbanity as assemblage', *Cambridge Journal of Regions, Economy and Society*, 7(1), pp. 119–136.

Färber, A. (2019) 'How does ANT help us to rethink the city and its promises?', in *The Routledge Companion to Actor-Network Theory*. Routledge, pp. 264–272.

Farias, I. (2009) 'Decentring the object of urban studies', in *Urban Assemblages: How Actor-Network Theory Changes Urban Studies*.

Farías, I. (2011) 'The politics of urban assemblages', *City*, 15(3–4), pp. 365–374.

Farias, I. and Bender, T. (2010) *Urban assemblages: How actor-network theory changes urban studies*. Routledge.

Farías, I. and Blok, A. (2017) 'STS in the City', in *The Handbook of Science and Technology Studies*. Cambridge: MIT Press.

Farmer, H. and Gabriel, M. (2020) *Innovation after Lockdown: Using innovation to build a more balanced, resilient economy*.

Flanagan, K. and Wilsdon, J. (2017) 'Powerhouse of science? Prospects and pitfalls of place-based science and innovation policies in northern England', in *Developing England's North: The Political Economy of the Northern Powerhouse*.

Florida, R. (2012) *The Rise of the Creative Class – Revisited*. New York: Basic Books.

Flowers, M. (2013) 'Beyond Open Data: The Data-Driven City', in *Open Data and the Future of Civic Innovation*.

Flyverbom, M., Deibert, R. and Matten, D. (2019) 'The Governance of Digital Technology, Big Data, and the Internet: New Roles and Responsibilities for Business', *Business and Society*, 58(1), pp. 3–19.

Foucault, M. (1980) 'Power/Knowledge: Selected Interviews & Other Writings'. Monoskop.

Foucault, M. (1996) 'What Is Critique?', in Schmidt, J. (ed.) *What Is Enlightenment?: Eighteenth-Century Answers and Twentieth-Century Questions*. University of California Press.

Fourcade, M. and Gordon, B. J. (2020) 'Learning Like a State: Statecraft in the Digital Age', *Journal of Law and Political Economy*, 1(1).

Fox, J. (2007) 'The uncertain relationship between transparency and accountability', *Development in Practice*, 17(4–5), pp. 663–671.

Fraser, D. (2016) *Whatever happened to Silicon Glen?, BBC News*.

Freeman, R. and Maybin, J. (2011) 'Documents, practices and policy', *Evidence and Policy*, pp. 155–170.

Freer, N. (2017) *Edinburgh's 'Holy Trinity of Tech'*.

Fujimura, J. H. (2003) 'Future imaginaries: Genome scientists as sociocultural entrepreneurs', *Genetic Nature/Culture: Anthropology and Science beyond the Two-Culture Divide*, (October), pp. 176–199.

Fuller, M. (2003) *Behind the Blip: Essays on the Culture of Software*. Autonomedia.

Geddes, P. (1904) 'Civics: as applied sociology', *The Sociological Review*, 1(1).

Gibbs, G. R. (2007) 'Thematic Coding and Categorizing In: Analyzing Qualitative Data'.

Gieryn, T. (1983) 'Boundary-Work and the Demarcation of Science from Non-Science: Strains and Interests in Professional Ideologies of Scientists', *American Sociological Review*, 48(6), pp. 781–795.

Gieryn, T. F. (1995) 'Boundaries of Science', in *Handbook of Science and Technology Studies*, pp. 41–58.

Gieryn, T. F. (1999) *Cultural Boundaries of Science: Credibility on the Line*. University of Chicago Press.

Gieryn, T. F. (2002) 'What buildings do', *Theory and Society*, 31(1), pp. 35–74.

Gieryn, T. F. (2006) 'City as truth-spot: Laboratories and field-sites in Urban Studies', *Social Studies of Science*, 36(1), pp. 5–38.

Gieryn, T. F. (2018) *Truth Spots – How Places make People Believe*. The University of Chicago Press.

Goddard, John; Vallance, P. (2013) 'Introduction', in *The university and the city*.

Goddard, J. (2009) *Reinventing the Civic University, Nesta Working Paper*.

Goddard, J. *et al.* (2014) 'Universities as anchor institutions in cities in a turbulent funding environment: Vulnerable institutions and vulnerable places in England', *Cambridge Journal of Regions, Economy and Society*, 7(2), pp. 307–325.

Goddard, J. (John B. and Vallance, P. (2013) *The university and the city*. Routledge.

Goddard, J. and Vallance, P. (2011) 'Universities and regional development', in Pike, A., Rodríguez-Pose, A., and Tomaney, J. (eds) *Handbook of Local and Regional Development*.

Goddard, J., Vallance, P. and Kempton, L. (2012) 'Chapter 2: The civic university: Connecting the global and the local', in *Universities, Cities and Regions: Loci for Knowledge and Innovation Creation*, pp. 43–63.

Godin, B. and Vinck, D. (2017) *Critical Studies of Innovation, Critical Studies of Innovation*.

Government, S. and COSLA (2019) *Local Governance Review – Developing and Testing Proposals*.

Graham, S. (2016) *Vertical – The City from Satellites to Bunkers*. London and New York: Verso.

Graham, S. and Marvin, S. (2001) *Splintering Urbanism: Networked Infrastructures, Technological Mobilities, and the Urban Condition*. London and New York: Routledge.

Granovetter, M. (1985) *Economic Action and Social Structure: The Problem of Embeddedness, American Journal of Sociology*.

Granovetter, M. S. and Swedberg, R. (2011) *The Sociology of Economic Life*. Routledge.

Green, B. (2021) 'Data Science as Political Action: Grounding Data Science in a Politics of Justice', *Journal of Social Computing*, 2(3), pp. 249–265.

Greenfield, A. (2013) *Against the Smart City*. New York: Do Publications.

Gregory, J. and Lock, S. J. (2008) 'The Evolution of "Public Understanding of Science": Public Engagement as a Tool of Science Policy in the UK', *Sociology Compass*, 2(4), pp. 1252–1265.

Gross, M. and McGoey, L. (2015) *Routledge international handbook of ignorance studies, Routledge International Handbook of Ignorance Studies*. Routledge.

Groves, C. (2017) 'Emptying the future: On the environmental politics of anticipation', *Futures*. Elsevier Ltd, 92, pp. 29–38.

Grundmann, R. (2017) 'The Problem of Expertise in Knowledge Societies', *Minerva*. Springer Netherlands, 55(1), pp. 25–48.

Guironnet, A. and Halbert, L. (2014) *The financialization of Urban Development Projects: Concepts, processes, and implications, Laboratoire Techniques Territoires et Sociétés*.

Guseva, A. and Mooney, H. (2018) 'Time, role of the past and varieties of fictional expectations: comments on Jens Beckert's Imagined futures', *Distinktion*.

Guy, Simon; Karvonen, A. (2011) 'Using Sociotechnical Methods: Researching Human-Technological Dynamics in the City', in *Understanding Social Research: Thinking Creatively About Method*. London: Sage Publications, pp. 120–133.

Hajer, M. A. (1995) *The Politics of Environmental Discourse: Ecological Modernization and the Policy Process, Social Forces*. Oxford University Press.

Hajer, M. and Versteeg, W. (2019) 'Imagining the post-fossil city: why is it so difficult to think of new possible worlds?', *Territory, Politics, Governance*, 7(2), pp. 122–134.

Hall, P. (2002) *Cities of Tomorrow: An Intellectual History of Urban Planning and Design since 1880*. Third edition. Wiley Blackwell.

Hall, P. (2014) *Cities of Tomorrow. An intellectual history of urban planning and design since 1880*. Fourth edition. Wiley Blackwell.

Halpern, O. (2014) *Beautiful Data – A History of Vision and Reason since 1945*. Duke University Press.

Harari, Y. N. (2017) 'Dataism Is Our New God', *New Perspectives Quarterly*, 34, pp. 36–43.

Haraway, D. (1988) 'Situated knowledges: The science question in feminism and the privilege of partial perspective', *Feminist Studies*, pp. 575–599.

Harding, Alan; Le Galès, P. (1997) 'Globalization, Urban Change and Urban Policies in Britain and France', in *The Limits of Globalization*.

Harrison, J. and Hoyler, M. (2014) 'Governing the new metropolis', *Urban Studies Journal Limited*, 51(11), pp. 2249–2266. d

Harvey, D. (1989) 'From Managerialism to Entrepreneurialism: The Transformation in Urban Governance in Late Capitalism', *Geografiska Annaler. Series B, Human Geography*, 71(1), p. 3.

Harvey, D. (2008) 'The Right to the City'. Verso.

Haug, P. (1986) 'US High Technology Multinationals and Silicon Glen', *Regional Studies*, 20(2), pp. 103–116.

Healey, P. and González, S. (2005) 'A Sociological Institutionalist Approach to the Study of Innovation in Governance Capacity', *Urban Studies*, 42(11), pp. 2055–2069.

Heidegger, M. (2010) *Being and Time*. State University of New York Press, Albany.

HESA (2020) *Buildings and spaces*.

HESA (2021a) *HE staff by HE provider and activity standard occupational classification 2014/15 to 2019/20*.

HESA (2021b) *Income and expenditure summary*.

Higher Education Statistics Agency (2021) *Where do HE students study?*

Hine, C. (2015) 'The E 3 Internet: Embedded , Embodied , Everyday Internet. What does ethnography offer for understanding.', *Ethnography for the Internet: Embedded, Embodied and Everyday.*, (2015), pp. 19–54.

HM Treasury (2015) *Summer budget 2015*.

HM Treasury (2016) 'Northern Powerhouse Strategy', *Policy Paper*, (November).

HM Treasury & Department for Business Innovation and Skills (2014) *Our plan for growth: science and innovation*.

Hollands, R. G. (2008) 'Will the real smart city please stand up? Intelligent, progressive or entrepreneurial?', *City*, 12(3), pp. 303–320.

Hong, A. H. and Sullivan, F. R. (2013) 'Towards an idea-centered, principle-base design to as creation approach support learning knowledge', *Educational Technology Research and Development,*. Thcor Soc, 57(5), pp. 613–627.

Hooley, T., Marriott, J. and Wellens, J. (2012) 'Online interviews and focus groups', in *What is Online Research?: Using the Internet for Social Science Research*. Bloomsbury.

Hoppe, R. (2005) 'Rethinking the science-policy nexus: From knowledge utilization and science technology studies to types of boundary arrangements', *Poiesis und Praxis*, 3(3), pp. 199–215.

House of Lords (2000) *Science and Technology – Third Report*.

Huggins, R. and Johnston, A. (2009) 'The economic and innovation contribution of universities: A regional perspective', *Environment and Planning C: Government and Policy*, 27(6), pp. 1088–1106.

Imam, J. (2021) 'Architects Dreaming of a Future with no Buildings', *New York Times*. A

Jacobs, J. (1961) 'The Death and Life Cities of Great American Cities'. New York: Vintage Books.

James, Nalita; Busher, H. (2012) 'Internet Interviewing', *The SAGE Handbook of Interview Research: The Complexity of the Craft*.

Jasanoff, S. (1990) 'Introduction – Rationalising Politics', in *The Fifth Branch – Science Advisers as Policymakers*. Harvard University Press.

Jasanoff, S. (2003a) 'Breaking the Waves in Science Studies: Comment on H.M. Collins and Robert Evans, "The Third Wave of Science Studies"', *Social Studies of Science*, 33(3), pp. 389–400.

Jasanoff, S. (2003b) 'Technologies of humility: Citizen participation in governing science', *Minerva*, 41, pp. 223–244.

Jasanoff, S. (2004) *States of Knowledge. The Co-Production of Science and Social Order*. Routledge.

Jasanoff, S. (2005) 'Judgment under Siege: The Three-Body Problem of Expert Legitimacy. | Harvard Kennedy School', in Maasen, Sabine; Weingart, P. (ed.) *Democratization of Expertise? Exploring Novel Forms of Scientific Advice in Political Decision-Making*. Springer.

Jasanoff, S. (2011) 'The Practices of Objectivity in Regulatory Science', *Social Knowledge in the Making*, pp. 307–337.

Jasanoff, S. (2012) *Science and Public Reason*. Routledge.

Jasanoff, S. (2015a) 'Future Imperfect', in *Dreamscapes of Modernity: Sociotechnical Imaginaries and the Fabrication of Power*, pp. 1–31. University of Chicago Press.

Jasanoff, S. (2015b) 'Future Imperfect – Science, Technology, and the Imginations of Modernity', in Jasanoff, Sheila; Kim, S.-H. (ed.) *Dreamscapes of Modernity*. University of Chicago Press, pp. 1–31.

Jasanoff, S. (2015c) 'Imagined and Invented Worlds', in *Dreamscapes of Modernity: Sociotechnical Imaginaries and the Fabrication of Power*.

Jasanoff, S. (2017) 'Science and Democracy', in Felt, Ulrike; Fouche, Rayvon; Miller, Clark; Smith-Doerr, L. (ed.) *The Handbook of Science and Technology Studies*.

Jasanoff, S. (2020a) 'Imagined worlds. The Politics of future-making in the twenty-first century', in *The Politics and Science of Prevision*. Abingdon, Oxon; New York: Routledge, pp. 27–44.

Jasanoff, S. (2020b) 'Imagined worlds – The Politics of Future-Making in the 21st Century', in *The Politics and Science of Prevision*. Abingdon, Oxon; New York: Routledge, pp. 27–44.

Jasanoff, S. and Kim, S.-H. (2009) 'Containing the Atom: Sociotechnical Imaginaries and Nuclear Power in the United States and South Korea', *Minerva*, 47(2), pp. 119–146.

Jasanoff, S. and Kim, S.-H. (2015) *Dreamscapes of Modernity: Sociotechnical Imaginaries and the Fabrication of Power*. University of Chicago Press.

Jasanoff, S., Kim, S. and Sperling, S. (2006) 'Sociotechnical Imaginaries and Science and Technology Policy: A Cross-National Comparison', pp. 1–21.

Jasanoff, S. and Simmet, H. R. (2021) 'Renewing the future: Excluded imaginaries in the global energy transition', *Energy Research and Social Science*. Elsevier Ltd, 80(January), p. 102205.

Jessop, B. (2002) 'Liberalism, Neoliberalism, and Urban Governance: A state-Theoretical Perspective', *Antipode*, pp. 452–472.

Jiménez, A. C. (2005) 'Changing scales and the scales of change: Ethnography and political economy in Antofagasta, Chile', *Critique of Anthropology*, pp. 157–176.

Johnson, J. (2015) 'One Nation Science'.

Joinson, A. N. and Paine, C. B. (2012) 'Self-disclosure, Privacy and the Internet', in *Oxford Handbook of Internet Psychology*, pp. 1–25.

Jones, A. *et al.* (2006) *Ideopolis: Knowledge City-Regions*.

Judd, D. and Laslo, D. (2013) 'The Regime Moment: The Briefbut StoriedCareer of Urban Regimes in American Cities', in *American Urban Politics in a Global Age*. Routledge.

Karvonen, A. (2020) 'Urban Techno-Politics: Knowing, Governing, and Imagining the City', *Science as Culture*, 29(3), pp. 417–424.

Karvonen, A. *et al.* (2021) 'The "New Urban Science": Towards the Interdisciplinary and Transdisciplinary Pursuit of Sustainable Transformations', *Urban Transformations*, 3(9), pp. 1–19.

Karvonen, A., Cugurullo, F. and Caprotti, F. (2019) *Inside Smart Cities. Place, Politics and Urban Innovation*. Routledge.

Karvonen, A. and van Heur, B. (2014) 'Urban laboratories: Experiments in reworking cities', *International Journal of Urban and Regional Research*, 38(2), pp. 379–392.

Karvonen, A., Martin, C. and Evans, J. (2018) 'University campuses as testbeds of smart urban innovation', in *Creating Smart Cities*. Abingdon, Oxon; New York, NY: Routledge.

Katz, B. and Wagner, J. (2014) '(PPT) The Rise of Innovation Districts: A New Geography of Innovation in America', *Brookings Institution*, (May), pp. 1–34.

Keynes, J. M. (1964) *The General Theory of Employment, Interest and Money, MacMillan*. London: Palgrave Macmillan UK.

Kingdon, J. W. (2014) *Agendas, Alternatives, and Public Policy.* Pearson Education Limited.

Kitchin, Rob; Coletta, Claudio; Evans, Leighton; Heaphy, Liam; Mac Donncha, D. (2017) *Smart cities, urban technocrats, epistemic communities and advocacy coalitions*.

Kitchin, R. (2014a) *The Data Revolution: Big Data, Open Data, Data Infrastructures & Their Consequences, The Data Revolution: Big Data, Open Data, Data Infrastructures & Their Consequences*.

Kitchin, R. (2014b) 'The real-time city? Big data and smart urbanism', *GeoJournal*, 79(1), pp. 1–14.

Kitchin, R. (2016) 'The ethics of smart cities and urban science', *Philosophical Transactions of the Royal Society A: Mathematical, Physical and Engineering Sciences*, 374(2083).

Kitchin, R. (2018) 'Reframing, Reimagining and Remaking Smart Cities', in *Creating Smart Cities*. Abingdon, Oxon; New York, NY: Routledge.

Kitchin, R. (2019) 'The Timescape of Smart Cities', *Annals of the American Association of Geographers*, pp. 1–16.

Kitchin, R., Cardullo, P. and Di Feliciantonio, C. (2019) 'Citizenship, Justice and the Right to the Smart City', in *The Right to the Smart City*.

Kitchin, R. and Dodge, M. (2011) *Code/Space. Software and Everyday Life*. MIT Press.

Klauser, F., Paasche, T. and Söderström, O. (2014) 'Michel foucault and the smart city: Power dynamics inherent in contemporary governing through code', *Environment and Planning D: Society and Space*, 32(5), pp. 869–885.

Knight, F. (1921) *Risk, Uncertainty and Profit, Sentry Press New York*. doi: 10.2307/1884757.

Knorr-Cetina, K. (1999) *Epistemic Cultures: How the Sciences Make Knowledge, Contemporary Sociology*.

Konrad, K. *et al.* (2017) 'Performing and Governing the Future in Science and Technology', in Miller, Clar; Felt, Ulrike; Fouché, Ray; Smith-Doerr, L. (ed.) *The Handbook of Science and Technology Studies*. Cambridge: MIT Press, pp. 465–493.

Kornberger, M. (2012) 'Governing the City. From Planning to Urban Strategy', *Theory, Culture & Society*, 29(2), pp. 84–106.

Kornberger, M. and Clegg, S. (2011) 'Strategy as performative practice', *Strategic Organization*, 9(2), pp. 136–162.

Kourtit, K., Nijkamp, P. and Arribas, D. (2012) 'Smart cities in perspective – a comparative European study by means of self-organizing maps', *Innovation*, 25(2), pp. 229–246.

Kummitha, R. K. R. and Crutzen, N. (2017) 'How do we understand smart cities? An evolutionary perspective', *Cities*, 67, pp. 43–52.

Kvale, S. (2007) *Doing Interviews*. London England EC1Y 1SP United Kingdom: SAGE Publications, Ltd.

Latour, B. (1990) 'Drawing Things Together', in Lynch, Michael; Woolgar, S. (ed.) *Representation in Scientific Practice*. MIT Press. A

Latour, B. (1996) *Aramis or the Love of Technology*. Harvard University Press.

Latour, B. (1999) 'Give Me a Laboratory and I Will Raise the World', in *The Science Studies Reader*. New York: Routledge.

Latour, B. (2004) 'Whose cosmos, which cosmopolitics? Comments on the Peace Terms of Ulrich Beck', *Kultur und Gerechtigkeit*, 10(3), pp. 43–57.

Latour, B. (2005) *Reassembling the Social – An Introduction to Actor-Network-Theory | bruno-latour.fr*. Oxford University Press.

Latour, B. and Hermant, E. (1998) *Paris: Invisible City*.

Law, J. (2002) 'Objects and Spaces', *Theory, Culture & Society*, 19(5).

Law, J. (2004) *After method: Mess in social science research, After Method: Mess in Social Science Research*.

Lawton-Smith, H. (2007) 'Universities, innovation, and territorial development: A review of the evidence', *Environment and Planning C: Government and Policy*, pp. 98–114.

Lee, N. and Clarke, S. (2019) 'Do low-skilled workers gain from high-tech employment growth? High-technology multipliers, employment and wages in Britain', *Research Policy*. Elsevier, 48(9).

Lefebvre, H. (1991) *The Production of Space*. Wiley.

Lefebvre, H. (1996) *Writings on Cities*. Blackwell Publishers.

Legg, S. (2011) 'Assemblage/apparatus: Using Deleuze and Foucault', *Area*, 43(2), pp. 128–133. doi: 10.1111/j.1475-4762.2011.01010.x.

Lente, H. van (1993) *Promising technology: The Dynamics of Expectations in Technological Developments*. Eburon.

Lessig, L. (2006) *Code. Version 2.0*. Basic Books.

Levitas, R. (2013) *Utopia as a Method. The Imaginary Reconstitution of Society, Palgrave Macmillan*. Palgrave MacMillan.

Light, J. (2003) 'From Warfare to Welfare Planning for the Atomic Age', in *From Warfare to Welfare*.

Lindner, C. and Meissner, M. (2019) 'The routledge companion to urban imaginaries', *The Routledge Companion to Urban Imaginaries*, pp. 1–466.

Loftman, P. and Nevin, B. (1996) 'Going for growth: Prestige projects in three British cities', *Urban Studies*, 33(6), pp. 991–1019.

Logan, M. (2020) *Scottish Technology Ecosystem Review*.

Lucas, R. E. (1972) 'Expectations and the Neutrality of Money', *Journal of Economic Theory*, 4, pp. 103–124.

Luque-Ayala, A. and Marvin, S. (2015) 'Developing a critical understanding of smart urbanism?', *Urban Studies*, 52(12), pp. 2105–2116.

Lynch, K. (2015) 'Control by numbers: new managerialism and ranking in higher education', *Critical Studies in Education*. Routledge, 56(2), pp. 190–207.

Lynch, K., Grummell, B. and Devine, D. (2012) *New managerialism in education: Commercialization, carelessness and gender, New Managerialism in Education: Commercialization, Carelessness and Gender*.

Lynch, M. (2003) 'The Creation of a College', in Anderson, R. D., Lynch, M., and Phillipson, N. (eds) *The University of Edinburgh: an illustrated history*. Edinburgh University Press, pp. 1–51.

Maasen, Sabine; Weingart, P. (2005) *Democratization of Expertise? Exploring Novel Forms of Scientific Advice in Political Decision-Making*. Springer.

MacLeod, G. and Goodwin, M. (1999) 'Space, scale and state strategy: Rethinking urban and regional governance', *Progress in Human Geography*, 23(4), pp. 503–527.

Mahoney, J. (2000) 'Path Dependence in Historical Sociology', *Theory and Society*, 29(4), pp. 507–548.

Manovich, L. (2000) *The Language of New Media*. MIT Press.

Mansfield, I. and Owen, G. (2020) *Visions of ARPA: Embracing Risk, Transforming Technology*.

Martin, R. *et al.* (2019) *The Economic Performance of Britain's Cities: Patterns , Processes and Policy Implications*.

Mason, J. (2002) *Qualitative Researching*. Sage Publications.

Masters, B. (2022) *Edinburgh bids for tech glory — again, Financial Times*.

Mayer-Schönberger, Viktor; Cukier, K. (2013) *Big Data: A Revolution That Will Transform How We Live, Work, and Think*. Boston: Houghton Mifflin Harcourt.

Mayer-Schönberger, V. and Cukier, K. (2013) *Big data: A revolution that will transform how we live, work, and think*.

McAra, L. (2019) 'Positive Contagion: What Universities are For', in *The Evergreen. A New Season in the North*. The Word Bank.

McCann, P. and Soete, L. (2020) *Place-based innovation for sustainability*.

McCrone, D. (2022) *Who Runs Edinburgh?* Edinburgh University Press.

McFall, L. and AWED (2020) *Closes and Opens: A History of Edinburgh's Futures*.

McFarlane, C. (2011) 'Assemblage and Critical Urbanism', *City*. Online, 15(2), pp. 204–224.

McFarlane, C. and Söderström, O. (2017) 'On alternative smart cities: From a technology-intensive to a knowledge-intensive smart urbanism', *City*. Taylor & Francis, 21(3–4), pp. 312–328.

McGuirk, P. M. (2000) 'Power and policy networks in urban governance: Local government and property-led regeneration in Dublin', *Urban Studies*, 37(4), pp. 651–672.

McLeod, M. (1983) '"Architecture or Revolution": Taylorism, Technocracy, and Social Change', *Art Journal*, 43(2), pp. 132–147.

McNeil, M. *et al.* (2017) 'Conceptualizing Imaginaries of Science, Technology and Society', in Felt, Ulrike; Fouché, Ray; Miller, Clark; Smith-Doerr, L. (ed.) *The Handbook of Science and Technology Studies*. Cambridge: MIT Press.

McQuillan, D. (2018) 'Data Science as Machinic Neoplatonism', *Philosophy and Technology*, 31(2), pp. 253–272.

Medina, E. (2011) *Cybernetic Revolutionaries – Technology and Politics in Allende's Chile*. MIT Press.

Medina, E. (2015) 'Rethinking algorithmic regulation', *Kybernetes*, 44(6–7), pp. 1005–1019.

Medvetz, T. (2012a) 'Gaining ground. The Rise of the Policy Expert', in *Think Tanks in America*.

Medvetz, T. (2012b) *Think tanks in America*. The University of Chicago Press.

Merton, R. K. (1973) *The sociology of science×: theoretical and empirical investigations*. University of Chicago Press.

Metzger, J. (2016) 'Cultivating torment: The cosmopolitics of more-than-human urban planning', *City*. Jonathan Metzger, 20(4), pp. 581–601.

Meyer, M. (2010) 'The rise of the knowledge broker', *Science Communication*, 32(1), pp. 118–127.

Miceli, M. and Posada, J. (2022) 'The Data-Production Dispositif', *Proceedings of the ACM on Human-Computer Interaction*, 6(CSCW2), pp. 1–37.

Miller, P. and Rose, N. (1990) 'Governing economic life', *Economy and Society*, 19(1), pp. 1–31.

Miller, P. and Rose, N. (2008) *Governing the Present*. Polity Press. doi: 10.1057/9781137445711_7.

Mintrom, M. (1997) 'Policy Entrepreneurs and the Diffusion of Innovation', *American Journal of Political Science*, 41(3), p. 738.

Mische, A. (2009) 'Projects and possibilities: Researching futures in action', *Sociological Forum*, 24(3), pp. 694–704.

Mische, A. (2014) 'Measuring futures in action: Projective grammars in the Rio+20 debates', *Theory and Society*, 43(3), pp. 437–464.

Mitchell, T. (1991) 'The Limits of the State⨯: Beyond Statist Approaches and Their Critics', *The American Political Science Review*, 85(1), pp. 77–96.

Mitchell, W. J. (1995) *City of Bits. Space, Place, and the Infobahn*. MIT Press.

Mitchell, W. J. (1999) *E-topia: Urban life, Jim – But Not as We Know it*. MIT Press.

Mitchell, W. J. (2003) *Me++. The Cyborg Self and the Networked City*. MIT Press.

Moisio, S. and Rossi, U. (2020) 'The start-up state: Governing urbanised capitalism', *Environment and Planning A*, 52(3), pp. 532–552.

Molotch, H. (1976) 'The City as a Growth Machine⨯: Toward a Political Economy of Place', *American Journal of Sociology*, 82(2), pp. 309–332.

De Monchaux, N. (2011) 'Cities and Cybernetics', in *Scales Of The Earth*, pp. 17–25.

Mora, L., Bolici, R. and Deakin, M. (2017) 'The First Two Decades of Smart-City Research: A Bibliometric Analysis', *Journal of Urban Technology*, 24(1), pp. 3–27.

Morozov, E. (2014) 'The Planning Machine – Project Cybersyn and the origins of the Big Data nation'.

Morozov, E. and Bria, F. (2018) *Rethinking the smart city, Rosa Luxemburg Stiftung*.

Mossberger, K. and Stoker, G. (2001) 'The evolution of urban regime theory: The challenge of conceptualization', *Urban Affairs Review*, 36(6), pp. 810–835.

Mrázek, R. (2002) *Engineers of Happy Land*. Princeton University Press.

Müller, M. (2015) 'Assemblages and Actor-networks: Rethinking Socio-material Power, Politics and Space', *Geography Compass*, 9(1), pp. 27–41.

Müller, M. and Schurr, C. (2016) 'Assemblage thinking and actor-network theory: conjunctions, disjunctions, cross-fertilisations', *Transactions of the Institute of British Geographers*, 41(3), pp. 217–229.

Mumford, E. (1992) 'CIAM urbanism after the Athens charter', *Planning Perspectives*, 7(4), pp. 391–417.

Münch, R. (2020) 'Academic Capitalism', in *Oxford Research Encyclopedia of Politics*.

Muth, J. F. (1961) *Rational Expectations and the Theory of Price Movements*, *Econometrica*.

Myers West, S. (2019) 'Data Capitalism: Redefining the Logics of Surveillance and Privacy', *Business & Society*, 58(1), pp. 20–41.

Neirotti, P. *et al.* (2014) 'Current trends in smart city initiatives: Some stylised facts', *Cities*, 38, pp. 25–36.

Nereim, V. (2022) 'MBS's $500 Billion Desert Dream Just Keeps Getting Weirder', *Bloomberg*.

Nijman, J. and Wei, Y. D. (2020) 'Urban inequalities in the 21st century economy', *Applied Geography*.

O'Brien, P. and Pike, A. (2015) 'City Deals, Decentralisation and the Governance of Local Infrastructure Funding and Financing in the UK', *National Institute Economic Review*, 233(1), pp. R14–R26.

O'Connor, H. *et al.* (2008) 'Internet-based Interviewing', in *The Sage Handbook of Online Research Methods*.

OECD (ed.) (2007) *Higher Education and Regions: Globally Competitive, Locally Engaged*. OECD Publishing.

Oomen, J., Hoffman, J. and Hajer, M. A. (2022) 'Techniques of futuring: On how imagined futures become socially performative', *European Journal of Social Theory*, 25(2), pp. 252–270.

Osborne, G. (2014) "We need a Northern powerhouse".

Park, C. and Allaby, M. (2013) *A Dictionary of Environment and Conservation*. Oxford University Press.

Park, R. E. (1936) 'Human Ecology', *American Journal of Sociology*, 42(1).

Parsons, T. (1951) *The social system*. London: Routledge.

Pasquale, F. and Sadowski, J. (2015) 'The spectrum of control: A social theory of the smart city', *First Monday*, 20(7), pp. 1–13.

Patton, M. Q. (2002) 'Designing Qualitative Studies', in *Qualitative Research & Evaluation Methods*.

Peck, J., Theodore, N. and Brenner, N. (2009) 'Neoliberal Urbanism: Models, Moments, Mutations', *SAIS Review*. Winter-Spring, 29(1), pp. 49–66.

Perman, R. (Journalist) (2018) *The rise and fall of the city of money: a financial history of Edinburgh*.

Perry, B. (2008) 'Academic Knowledge and Urban Development', in *Knowledge-Based Urban Development*. IGI Global, pp. 21–41.

Perry, D. C. and Wiewel, W. (2005) *The university as urban developer: Case studies and analysis: Case studies and analysis, The University as Urban Developer: Case Studies and Analysis: Case Studies and Analysis*. Routledge.

Pfotenhauer, S. *et al.* (2022) 'The politics of scaling', *Social Studies of Science*, 52(1), pp. 3–34.

Pfotenhauer, S. and Jasanoff, S. (2017a) 'Panacea or diagnosis? Imaginaries of innovation and the "MIT model" in three political cultures', *Social Studies of Science*, 47(6), pp. 783–810.

Pfotenhauer, S. and Jasanoff, S. (2017b) 'Travelling Imaginaries – The Practice turn in innovation policy and the global circulation of innovation models', in *The Routledge Handbook of the Political Economy of Science*. Taylor & Francis.

Pfotenhauer, S. M. and Juhl, J. (2017) 'Innovation and the political state: beyond the myth of technologies and markets', in Godin, Benoit; Vinck, D. (ed.) *Critical Studies of Innovation*. Edward Elgar Publishing, pp. 68–94.

Pfotenhauer, S. M., Juhl, J. and Aarden, E. (2019) 'Challenging the "deficit model" of innovation: Framing policy issues under the innovation imperative', *Research Policy*. Elsevier, 48(4), pp. 895–904.

Phillipson, N. (2003) 'The Making of an Enlightened University', in *The University of Edinburgh: an illustrated history*. Edinburgh University Press.

Pickering, A. (2010) *The Cybernetic Brain: Sketches of Another Future*. The University of Chicago Press.

Picon, A. (2015) *Smart cities: A spatialised intelligence, Smart Cities: A Spatialised Intelligence*.

Pollio, A. (2016) 'Technologies of austerity urbanism: the "smart city" agenda in Italy (2011–2013)', *Urban Geography*. Routledge, 37(4), pp. 514–534.

Pollock, N. and Williams, R. (2018) *How Industry Analysts Shape the Digital Future*. Oxford University Press.

Power, D. and Malmberg, A. (2008) 'The contribution of universities to innovation and economic development: In what sense a regional problem', *Cambridge Journal of Regions, Economy and Society*, 1(2), pp. 233–245.

Rammert, W. *et al.* (2018) *Innovation society today: Perspectives, fields, and cases, Innovation Society Today: Perspectives, Fields, and Cases*.

Rammstedt, O. (1975) 'Alltagsbewußtsein von Zeit', *Kölner Zeitschrift für Soziologie und Sozialpsychologie*, 27(1).

Rhodes, R. A. W. (1997) *Understanding governance: Policy networks, Governance, Reflexivity, and Accountability*. Open University Press.

Riach, A. (2017) *The Revolutionary: Reconsidering the work of Hugh MacDiarmid, The National*. A

Richards, D. (1996) 'Elite Interviewing: Approaches and Pitfalls', *Politics*, 16(3), pp. 199–204.

Riese, J. (2018) 'What is "access" in the context of qualitative research?', *Qualitative Research*, (1757).

Rist, G. (2007) 'Development as a Buzzword', *Source: Development in Practice*, 17(4), pp. 485–491.

Rittel, H. W. and Webber, M. M. (1973) 'Dilemmas in a General Theory of Planning', *Policy Sciences*, 4, pp. 155–169.

Rosa, H. (2013) *Social acceleration: A New Theory of Modernity*. Columbia University Press.

Rose, G. (2019) 'Smart urban: imaginary, interiority, intelligence', in *The Routledge Companion to Urban Imaginaries*.

Rydin, Y. (2013) 'Using Actor-Network Theory to understand planning practice: Exploring relationships between actants in regulating low-carbon commercial development', *Planning Theory*, 12(1), pp. 23–45.

Sadowski, J. (2020) 'Who owns the future city? Phases of technological urbanism and shifts in sovereignty', *Urban Studies*, (February).

Sadowski, J. and Bendor, R. (2018) 'Selling Smartness', *Science, Technology, & Human Values*, p. 016224391880606.

Sager, T. (2011) 'Neo-liberal urban planning policies: A literature survey 1990–2010', *Progress in Planning*, 76(4), pp. 147–199.

Schiølin, K. (2020) 'Revolutionary dreams: Future essentialism and the sociotechnical imaginary of the fourth industrial revolution in Denmark', *Social Studies of Science*, 50(4), pp. 542–566.

Schneider, C. and Lösch, A. (2019) 'Visions in assemblages: Future-making and governance in FabLabs', *Futures*. Elsevier Ltd, 109(August 2018), pp. 203–212.

Schwab, K. (2016) *The Fourth Industrial Revolution: what it means, how to respond*.

Schwartz, J. M. (2014) 'Resisting the Exploitation of Contingent Faculty Labor in the Neoliberal University: The Challenge of Building Solidarity between Tenured and Non-Tenured Faculty', *New Political Science*, 36(4), pp. 504–522.

Scoones, I. (2007) 'Sustainability', *Development in Practice*, 17(4–5), pp. 589–596.

Scott, J. C. (1998) *Seeing Like a State. How certain schemes to improve the human condition have failed, Seeing Like a State*. Yale University Press.

Scottish Government (2017) 'Realising Scotland's full potential in a digital world: a digital strategy for Scotland', p. 19.

Scottish Government (2021a) 'A Changing Nation: How Scotland will Thrive in a Digital World'.

Scottish Government (2021b) *National Performance Framework*.

Scottish Government (2021c) *Scotland's Artificial Intelligence Strategy*.

Scottish Government (2022a) 'A stronger economy with independence'.

Scottish Government (2022b) *Building a tech nation*.

Sennett, R. (2006) *The open city*. Available at: https://urbanage.lsecities.net/essays/the-open-city.

Sewell, W. H. (2008) 'The temporalities of capitalism', *Socio-Economic Review*, 6(3), pp. 517–537.

Shackle, G. (1979) *Imagination and the Nature of Choice*. Edinburgh University Press.

Shakespeare, S. (2013) *An Independent Review of Public Sector Information*.

Simmel, G. (1903) 'The Metropolis and Mental Life'. Oxford and Malden.

Simmel, G. (2004) *The philosophy of money*. Routledge.

Singh, G., Kahlon, S. and Chandel, V. B. S. (2019) 'Political Discourse and the Planned City: Nehru's Projection and Appropriation of Chandigarh, the Capital of Punjab', *Annals of the American Association of Geographers*, 109(4), pp. 1226–1239.

Skotnicki, T. and Nielsen, K. (2021) 'Toward a theory of alienation: futurelessness in financial capitalism', *Theory and Society*, 50(6), pp. 837–865.

Slaughter, Sheila; Rhoades, G. (2004) *Academic Capitalism and the New Economy – Markets, State, and Higher Education*, *The John Hopkins University Press*.

Slaughter, S. and Rhoades (2000) 'The Neo-Liberal University', *New Labor Forum*, 6(6), pp. 73–79.

Smallman, M., Lock, S. J. and Miller, S. (2020) 'United Kingdom: The developing relationship between science and society', *Communicating Science: A Global Perspective*, pp. 931–957.

Smith, E. (2015) 'Corporate Imaginaries of Biotechnology and Global Governanc', in *Dreamscapes of Modernity*. University Press Scholarship.

Snieska, V. and Zykiene, I. (2015) 'City attractiveness for investment: characteristics and underlying factors', *Procedia-Social and Behavioral Sciences*, 213, pp. 48–54.

Söderström, O., Paasche, T. and Klauser, F. (2014) 'Smart cities as corporate storytelling', *City*, 18(3), pp. 307–320.

Sørensen, K. H. and Traweek, S. (2021) *Navigating Universities, University Studies, and This Book, Questing Excellence in Academia*.
Srnicek, N. (2016) 'Introduction', in *Platform Capitalism*.
Star, S. L. and Griesemer, James R. (1989) 'Institutional Ecology, "Translations" and Boundary Objects: Amateurs and Professionals in Berkeley's Museum of Vertebrate Zoology, 1907–39', *Social Studies of Science*, 19(3), pp. 387–420.
Star, S. L. and Griesemer, James R (1989) 'Institutional Ecology, "Translations" and Boundary Objects: Amateurs and Professionals in Berkeley's Museum of Vertebrate Zoology , 1907–39 Author (s): Susan Leigh Star and James R . Griesemer Source: Social Studies of Science , Vol . 19 , No . 3', *Social Studies of Science*, 19(3), pp. 387–420.
Stengers, I. (2005) 'A Cosmopolitical Proposal', in Latour, B. and Weibel, P. (eds) *Making Things Public: Atmospheres of Democracy*. MIT Press.
Stephens, N., Atkinson, P. and Glasner, P. (2013) 'Science as Culture Institutional Imaginaries of Publics in Stem Cell Banking: The Cases of the UK and Spain'.
Stevenson, D. (2003) *Cities and Urban Cultures*. Open University Press.
Stoker, G. (2011) 'Was local governance such a good idea? A global comparative perspective', *Public Administration*, 89(1), pp. 15–31.
Stone, C. N. (1989) *Regime politics: governing Atlanta, 1946–1988*. University Press of Kansas.
Stone, C. N. (1993) 'Urban regimes and the capacity to govern', *Journal of Urban Affairs*, 15(1), p. 1.
Stone, C. N. (2015) 'Reflections on Regime Politics: From Governing Coalition to Urban Political Order', *Urban Affairs Review*, 51(1), pp. 101–137.
Sutherland, E. (2020) 'Edinburgh: Data Capital of Europe', *SSRN Electronic Journal*, pp. 1–16.
Svetlova, E. (2022) 'AI meets narratives: the state and future of research on expectation formation in economics and sociology', *Socio-Economic Review*, 20(2), pp. 841–861.
Taylor Buck, N. and While, A. (2017) 'Competitive urbanism and the limits to smart city innovation: The UK Future Cities initiative', *Urban Studies*, 54(2), pp. 501–519.
Taylor, C. (2003) *Modern social imaginaries*. Duke University Press.
The City of Edinburgh Council (2018) *Edinburgh and South-East Scotland City Region Deal*.
The City of Edinburgh Council (2020a) *City Region Deal Benefits Realisation Plan*.

The City of Edinburgh Council (2020b) *Council extends partnership with CGI as it aims to become one of the world's 'smartest cities''*.
The City of Edinburgh Council (2020c) *Edinburgh by Numbers*.
The City of Edinburgh Council (2021) *Edinburgh and South East Scotland Regional Prosperity Framework (2021–2041)*.
The Scottish Government (2015) *Scotland's Economic Strategy*.
The University of Edinburgh (2020) *Community Plan 2020–25*.
The University of Edinburgh (2021a) *Annual Report and Accounts for the Year to 31 July 2021*.
The University of Edinburgh (2021b) *Opening of Edinburgh University, 1583 – Our History*.
The University of Edinburgh Estates Department (2017) *Estates Vision 2017 – 2027*.
Thomas Kuhn (1962) *The structure of scientific revolutions*. University of Chicago Press.
Thorp, H. and Goldstein, B. (2013) *Engines of Innovation: The Entrepreneurial University in the Twenty-First Century*.
Thrift, N. (2014) 'The "sentient" city and what it may portend', *Big Data & Society*, 1(1), p. 2053951714532241.
Tönnies, F. (2012) *Gemeinschaft und Gesellschaft, Studien zu Gemeinschaft und Gesellschaft*. Wiesbaden: VS Verlag für Sozialwissenschaften.
Tornatzky, L., Waugaman, P. and O'Gray, D. (2002) *Innovation U: New University Roles in a Knowledge Economy, Research Triangle Park, N.C.*
Townsend, A. (2015) 'Cities of data: Examining the new urban science', *Public Culture*, 27(2), pp. 201–212.
Townsend, A. M. (2013) 'Smart Cities – Big data, civic hackers, and the quest for a new utopia'. W.W. Norton & Company.
Tuchman, G. (2009) *Wannabe U. Inside the Corporate University*.
UK Government (2015) 'Business Secretary sharpens the UK's innovation expertise with regional research and development audit – GOV.UK'.
UK Government (2017) *Enabling a World-Leading Regional Digital Economy through Data Driven Innovation*.
UK Government (2020) 'UK Research and Development Roadmap', (July).
UK Government (2021) *Build Back Better our plan for growth Build Back Better: our plan for growth*.
UN-Habitat (2022) *Envisaging the Future of Cities, World City Report*.
UN (2018) *World Urbanization Prospects, Demographic Research*.
UNEP – UN Environment Programme (2022) *Cities and climate change*.

Urry, J. (2016) *What is the future?* Wiley.

Uyarra, E. (2010) 'Conceptualizing the regional roles of universities, implications and contradictions', *European Planning Studies*. Elvira Uyarra, 18(8), pp. 1227–1246.

Vanolo, A. (2014) 'Smartmentality: The Smart City as Disciplinary Strategy', *Urban Studies*, 51(5), pp. 883–898.

Venugopal, R. (2015) 'Neoliberalism as concept', *Economy and Society*, 44(2), pp. 165–187.

Virilio, P. and Lotringer, S. (2007) *Pure War*. Semiotexte.

Waite, D. *et al.* (2018) *The emergence and evolution of City Deals in Scotland*.

Wakeman, R. (2003) 'Dreaming the New Atlantis: Science and the Planning of Technopolis, 1955–1985', in *Osiris: Science and the City*, pp. 255–270.

Ward, M. (2018) *City Deals*. Available at: https://www.gov.uk/government/collections/city-deals.

Weber, M. (1958) *The City*. The Free Press.

Weber, R. (2010) 'Selling City Futures: The Financialisation of Urban Redevelopment Policy', *Economic Geography*, 86(3), pp. 251–274.

Weingart, P. (1999) 'Scientific expertise and political accountability: Paradoxes of science in politics', *Science and Public Policy*, 26(3), pp. 151–161. doi: 10.3152/147154399781782437.

Wenger, Andreas; Jasper, Ursula; Dunn Cavelty, M. (2020) *The Politics and Science of Prevision – Governing and Probing the Future*. Routledge.

Whyte, W. H. (2016) *Redbrick. A social and architectural history of Britain's civic universities*. Oxford University Press.

Wiener, N. (2019) *Cybernetics or Control and Communication in the Animal and the Machine*. MIT Press.

Wiig, A. (2015) 'IBM's smart city as techno-utopian policy mobility', *City*, 19(2–3), pp. 258–273.

Wildayvsky, A. (1973) 'If planning is everything, maybe it's nothing', in *Policy Sciences*, pp. 127–153.

Winner, L. (1986) *The Whale and the Reactor*. The University of Chicago Press.

Wynne, B. (1996) 'May the Sheep Safely Graze? A Reflexive View of the Expert–Lay Knowledge Divide', in Lash, S.; Szerszynski, B; Winne, B. (ed.) *Risk, Environment and Modernity: Towards a New Ecology*. London: SAGE Publications Ltd, pp. 44–83.

Yigitcanlar, T. (2011) 'Position paper: redefining knowledge-based urban development', *International Journal of Knowledge-Based Development*, 2(4), pp. 340–356.

Yigitcanlar, T., Velibeyoglu, K. and Baum, S. (2008) *Knowledge-based urban development: Planning and applications in the information era, Knowledge-Based Urban Development: Planning and Applications in the Information Era*.

Yin, R. K. (2008) *Case Study Research: Design and Methods*.

Zanella, A. *et al.* (2014) 'Internet of things for smart cities', *IEEE Internet of Things Journal*. IEEE, 1(1), pp. 22–32.

Zehner, N. (2022) 'The Civic University as Key Agent in the Production of Urban Space', in Currie, M., Knox, J., and McGregor, C. (eds) *Data Justice and the Right to the City*. Edinburgh University Press.

Zeiderman, A., Dawson, K. and Palmer, C. (2021) 'Urban Futures , Past and Present Geography and Environment Discussion Paper Series February 2021 Editorial Board', (23).

Zuboff, S. (2015) 'Big other: Surveillance capitalism and the prospects of an information civilization', *Journal of Information Technology*. Nature Publishing Group, 30(1), pp. 75–89.

Zukin, S. (2020a) 'Imagining Innovation', in *The Innovation Complex: Cities, Tech, and the New Economy*, pp. 1–26.

Zukin, S. (2020b) 'Pipelines: Talent, Meritocracy, and Academic Capitalism', in *The Innovation Complex: Cities, Tech, and the New Economy*.

Zukin, S. (2020c) 'Seeing like a city: how tech became urban.', *Theory and society*. Theory and Society, pp. 1–24.

Zukin, S. (2020d) *The Innovation Complex, The Innovation Complex*. Oxford University Press.

Zukin, S. (2021) 'Planetary Silicon Valley: Deconstructing New York's innovation complex', *Urban Studies*. SAGE Publications Ltd, 58(1), pp. 3–35.

Appendix

Figure 38: Documents | Source: Own illustration

Total	University of Edinburgh	Edinburgh City Council	UK Government	Other	Scottish Government
48	6	7	9	10	16
	Strategic Plan 2016	City Region Deal 2018	BEIS Science and Innovation Audit Wave 1 Summaries	CBI Report "Stronger Together" 2009	Data Vision 2014
	Strategic Plan 2020	Economy Strategy 2018	Digital Strategy 2017	KPMG Report "UK City Deals" 2014	Digital Strategy 2011
	Community Plan 2020-25	Local Development Plan 2016	Technology Innovation Strategy 2019	Lord Sainsbury Report "Review of the UK Government's Science and Innovation Policies" 2007	Digital Strategy 2017
	Estates Vision 2017-2027	Choices for Cities 2030 Plan	Industrial Strategy 2017	Scottish City Alliances "Smart Cities Scotland Blueprint" 2016	Economic Strategy 2007
	Annual Report 2020	Sustainable Edinburgh 2020 Vision	SIA Report Edinburgh City Region Deal 2017	World Economic Forum "The Fourth Industrial Revolution"	Economic Strategy 2011
	Annual Report 2021	Tourism Strategy 2030	Scotland Office Vision	OECD Report "Supporting Investment in Knowledge Capital, Growth and Innovation"	Economic Strategy 2015
		Regional Prosperity Framework 2021	White Paper "Realising Every Place's Potential" 2010	Nesta report "Innovation after Lockdown" 2020	Government Programme 2019-2020
			Consultation Paper "Unlocking Growth" 2011	Nesta report "Is Scotland getting Innovation Right?"	Inclusive Growth Outcomes Framework 2019

Figure 39: List of Interviewees | Source: Own illustration

Synonym	Location	Org. Type	Profession	Date (Interview 1)	Date (Interview 2)
Interview 1	Edinburgh	Local government	Council Leader	07.11.2019	
Interview 2	Edinburgh	Higher education	Programme Manager	11.11.2019	
Interview 3	Glasgow	Higher education	Professor	20.11.2019	
Interview 4	Online	Third sector	CEO	08.09.2020	
Interview 5	Edinburgh	Third sector	CEO	22.01.2020	
Interview 6	Edinburgh	Third sector	Chairman	23.01.2020	
Interview 7	Edinburgh	Higher education	CEO	10.02.2020	
Interview 8	Edinburgh	Local government	Programme Manager	20.02.2020	
Interview 9	Edinburgh	Business	Policy Director	24.02.2020	
Interview 10	Edinburgh	Higher education	Programme Manager	25.02.2020	15.05.2020
Interview 11	Edinburgh	Business	Director	03.03.2020	
Interview 12	Online	Business	CEO	12.03.2020	
Interview 13	Online	National government	Lead Policy Officer	23.03.2020	
Interview 14	Online	Third sector	CEO	31.03.2020	
Interview 15	Online	Local government	Programme Manager	01.04.2020	
Interview 16	Online	Higher education	Programme Manager	07.04.2020	
Interview 17	Online	Business	CEO	08.04.2020	
Interview 18	Online	International organisation	Director	08.04.2020	
Interview 19	Online	Higher education	Professor	17.04.2020	
Interview 20	Online	National government	Deputy Director	21.04.2020	
Interview 21	Online	Higher education	COO	23.04.2020	
Interview 22	Online	National government	Desk Officer	24.04.2020	
Interview 23	Online	Business	Director	30.04.2020	
Interview 24	Online	Higher education	Vice Chancellor	11.05.2020	
Interview 25	Online	Higher education	Assistant Principal	13.05.2020	
Interview 26	Online	Higher education	Programme Manager	15.05.2020	
Interview 27	Online	Local government	Former Council Leader	18.05.2020	
Interview 28	Online	Higher education	Professor	20.05.2020	
Interview 29	Online	Higher edu-	Programme	26.05.2020	

		cation	Manager		
Interview 30	Online	Local government	Council Leader	27.05.2020	
Interview 31	Online	Third sector	CEO	02.06.2020	
Interview 32	Online	Third sector	Programme Director	12.06.2020	
Interview 33	Online	Third sector	Desk Officer	16.06.2020	
Interview 34	Online	Higher education	Programme Manager	17.06.2020	
Interview 35	Online	Local government	Council Leader	07.07.2020	
Interview 36	Online	Local government	Lord Provost	14.07.2020	
Interview 37	Online	Third sector	CEO	30.07.2020	
Interview 38	Online	Third sector	CEO	25.08.2020	
Interview 39	Online	National government	Desk Officer	25.08.2020	
Interview 40	Online	Third sector	CEO	27.08.2020	
Interview 41	Online	National government	Deputy Director	28.08.2020	
Interview 42	Online	National government	Lead Policy Officer	10.09.2020	
Interview 43	Online	Local government	CEO	11.11.2020	
Interview 44	Online	Higher education	Programme Manager	27.11.2020	
Interview 45	Online	Local government	CEO	07.12.2020	
Interview 46	Online	Higher education	Senior Vice Principal	08.12.2020	
Interview 47	Online	National government	Policy Manager	08.12.2020	
Interview 48	Online	Local government	CEO	10.12.2020	
Interview 49	Online	Higher education	Programme Manager	27.05.2021	
Interview 50	Online	Higher education	Professor	02.06.2021	

Figure 40: List of Observations | Source: Own illustration

Type	Location	Date
Town hall meeting	Edinburgh	22.01.2020
Industry conference	Oxford	05.02.2020
Industry conference	London	13.02.2020
Policy conference	Online	29.04.2020
Policy conference	Online	06.11.2020
Policy conference	Online	23.03.2021
Policy conference	Online	10.05.2021
Policy conference	Online	03.12.2021
Policy conference	Online	14.12.2021
Policy conference	Online	26.09.2022

Figure 41: Beer's ontological spaces | Source: Cybernetics and Management (Beer, 1959, p. 18)

SYSTEMS	*Simple*	*Complex*	*Exceedingly complex*
Deterministic	Window catch	Electronic digital computer	EMPTY
	Billiards	Planetary system	
	Machine-shop lay-out	Automation	
Probabilistic	Penny tossing	Stockholding	The economy
	Jellyfish movements	Conditioned reflexes	The brain
	Statistical quality control	Industrial profitability	THE COMPANY

Figure 42: Memo template | Source: Own illustration

Memo:		
Task	**Emerging thoughts**	**Evidence from data**
"Define the properties" of this category		
"Specify the conditions when the category develops, is maintained, and changes"		
"Delineate its consequences and relationships with other categories"		
"May also include the researcher's experiential struggles in making sense of data"		